T0140834

Anne Koch

# Determining input-output properties of linear time-invariant systems from data

Logos Verlag Berlin

Bibliographic information published by the Deutsche Nationalbibliothek

The Deutsche Nationalbibliothek lists this publication in the Deutsche Nationalbibliografie; detailed bibliographic data are available on the Internet at http://dnb.d-nb.de

D93



ISBN 978-3-8325-5446-0

Logos Verlag Berlin GmbH
Georg-Knorr-Str. 4, Geb. 10,
D-12681 Berlin
Germany

Tel.: +49 (0)30 / 42 85 10 90
Fax: +49 (0)30 / 42 85 10 92
http://www.logos-verlag.de

# Determining input-output properties of linear time-invariant systems from data

Von der Fakultät Konstruktions-, Produktions- und Fahrzeugtechnik
der Universität Stuttgart zur Erlangung der Würde eines
Doktor-Ingenieurs (Dr.-Ing.) genehmigte Abhandlung

Vorgelegt von

Anne Koch geb. Romer

aus Biberach an der Riß

Hauptberichter: Prof. Dr.-Ing. Frank Allgöwer
Mitberichter: Prof. Maria Prandini, Ph.D.
Prof. Dr. sc. Sebastian Trimpe

Tag der mündlichen Prüfung: 25.11.2021

Institut für Systemtheorie und Regelungstechnik
Universität Stuttgart
2022

# Acknowledgements

The results presented in this thesis are the outcome of my research activity at the Institute for Systems Theory and Automatic Control (IST) at the University of Stuttgart. There are numerous people who supported me throughout this time, and to whom I would like to express my heartfelt appreciation.

First of all, I want to express my deepest gratitude to my supervisor Prof. Frank Allgöwer for his continued support and trust throughout my time at the institute, providing me with countless opportunities and a lot of freedom in my research, as well as for creating such a unique and fruitful research environment. I also want to thank Prof. Maria Prandini (Politecnico di Milano) and Prof. Sebastian Trimpe (RWTH Aachen) for their interest in my work, for the valuable comments, and for being members of my doctoral examination committee.

Furthermore, I am grateful to Julian Berberich, Jan Maximilian Montenbruck, and Johannes Köhler for the close collaboration which had a great influence on my research. In addition, I had the pleasure to collaborate with many more outstanding researchers and students from all over the world. This includes Amr Alanwar, Prof. Sofie Haesaert, Prof. Karl Henrik Johansson, Matthias Lorenzen, Matias I. Müller, Patricia Pauli, Daniel Persson, Prof. Cristian R. Rojas, Prof. Carsten W. Scherer, Lukas Schwenkel, Miel Sharf, Prof. Roland Tóth, Chris Verhoek, Nils Wieler, and Prof. Daniel Zelazo. I am grateful to each one of them. I also want to express my gratitude to Prof. Bo Wahlberg for hosting me for three months at the KTH in Stockholm for an inspiring research stay, and I additionally want to acknowledge the Cluster of Excellence 'Data-integrated Simulation Science' (SimTech) and the International Max Planck Research School for Intelligent Systems (IMPRS-IS) for their support.

Moreover, I want to thank all of my current and former colleagues at the University of Stuttgart not only for the collaborative atmosphere, but also for the enjoyable lunch and coffee breaks, soccer and table soccer tournaments, and the many memorable conference travels.

Finally, I am indebted to my family and friends for their support, their understanding and patience, for jointly celebrating victories and sharing sorrows. In particular, I am incredibly grateful to my parents, who always and unconditionally supported me in any way imaginable. I also want to express my deepest gratitude to Markus, who was with me on this journey and whose support throughout the past years was essential for the success of this thesis. Lastly, I do not want to miss mentioning my little son Noah, who let me know with increasingly strong kicks in my belly that it was about time to finalize this thesis.

Friedrichshafen, December 2021
Anne Koch

# Table of Contents

# Notation

In the following, we list the main symbols and acronyms used in this thesis. Additional notation is defined in the corresponding sections.

### Abbreviations and acronyms

| | |
|---|---|
| FIR | finite impulse response |
| IQC | integral quadratic constraint |
| i.i.d. | independent and identically distributed |
| LFT | linear fractional transformation |
| LMI | linear matrix inequality |
| LQR | linear quadratic regulator |
| LTI | linear time-invariant |
| MIMO | multiple-input multiple-output |
| PN | positive-negative |
| SDP | semidefinite program |
| SISO | single-input single-output |
| SNR | signal-to-noise ratio |
| s.t. | such that |
| w.r.t. | with respect to |

### Real numbers, complex numbers, and sets

| | |
|---|---|
| $\mathbb{N}$ | Set of nonnegative integers. |
| $\mathbb{N}_+$ | Set of positive integers. |
| $\mathbb{Z}$ | Set of integers. |
| $\mathbb{R}$ | Set of reals. |
| $\mathbb{R}_{\geq 0}$ | Set of nonnegative reals. |
| $\mathbb{S}^{n-1}$ | $(n-1)$-dimensional unit sphere, i.e., $\mathbb{S}^{n-1} = \{x \in \mathbb{R}^n \mid \|x\| = 1\}$. |
| $\mathbb{C}$ | Set of complex numbers. |
| $\mathbf{i}$ | Imaginary unit, i.e., $\mathbf{i}^2 = -1$. |
| $\lvert z \rvert$ | Absolute value of $z \in \mathbb{C}$. |
| $l_2^n$ | Hilbert space consisting of square summable vector-valued sequences $x = (x_0, x_1, \dots)$, $x_k \in \mathbb{R}^n$, where $\|x\|^2 = \sum_{k=0}^{\infty} x_k^\top x_k < \infty$. |
| $l_{2,c}^n$ | Hilbert space consisting of square integrable functions $x : \mathbb{R}_{\geq 0} \to \mathbb{R}^n$, where $\|x\|^2 = \int_{t=0}^{\infty} x(t)^\top x(t)\, \mathrm{d}t < \infty$. |
| $\mathcal{RL}_\infty^{p\times q}$ | Set of all real-rational transfer matrices of dimension $p \times q$ essentially bounded on the unit circle. |
| $\mathcal{RH}_\infty^{p\times q}$ | Set of all transfer matrices in $\mathcal{RL}_\infty^{p\times q}$ that are analytic outside the unit circle. |

### Vectors, matrices, and norms

| | |
|---|---|
| $I_n$ | Identity matrix with dimension $n \times n$. |
| $0_{n\times m}$ | Matrix of zeros with dimension $n \times m$. |
| $A^\top$ | Transpose of the matrix $A \in \mathbb{R}^{n\times m}$. |
| $\ker A$ | Kernel of the matrix $A$. |
| $A^\perp$ | Orthogonal complement of $A$, i.e., a matrix which contains the column vectors spanning the kernel of $A$. |
| $x$ | Denotes either a sequence $\{x_k\}_{k=0}^{N-1}$ itself or the stacked vector containing its components. |
| $\|x_k\|$ | Euclidean norm of $x_k \in \mathbb{R}^n$. |
| $\|x\|^2$ | For a sequence $\{x_k\}_{k=0}^{N-1}$, we denote $\|x\|^2 = \sum_{k=0}^{N-1} \|x_k\|^2$. |

| | |
|---|---|
| $\lambda_i(Q)$, $i = 1, \ldots, n$ | $n$ eigenvalues of a symmetric matrix $Q \in \mathbb{R}^{n \times n}$, where $\lambda_1 (= \lambda_{\max}) \geq \lambda_2 \geq \cdots \geq \lambda_n (= \lambda_{\min})$. |
| $\lambda_i(P, Q)$, $i = 1, \ldots, n$ | $n$ generalized eigenvalues of symmetric matrices $P, Q \in \mathbb{R}^{n \times n}$, i.e., constants $\lambda_i \in \mathbb{R}$ with $P - \lambda_i Q$ singular, where $\lambda_1 (= \lambda_{\max}) \geq \lambda_2 \geq \cdots \geq \lambda_n (= \lambda_{\min})$. |
| $A \succ 0$ ($A \succeq 0$) | Matrix $A \in \mathbb{R}^{n \times n}$ is positive definite (positive semidefinite), i.e., $A = A^\top$ and $x^\top A x > 0$ ($x^\top A x \geq 0$) for all $x \in \mathbb{R}^n$ with $x \neq 0$. |
| $A \prec 0$ ($A \preceq 0$) | Matrix $A \in \mathbb{R}^{n \times n}$ is negative definite (negative semidefinite), i.e., the matrix $-A$ is positive definite (positive semidefinite). |
| $\|A\|$ | Matrix norm of a matrix $A \in \mathbb{R}^{m \times n}$ induced by the Euclidean norm, i.e., $\|A\| = \max_{x \in \mathbb{R}^n, \|x\|=1} \|Ax\| = \sqrt{\lambda_{\max}(A^\top A)}$. |

**Operators, functions, and derivatives**

| | |
|---|---|
| $B \otimes C$ | Kronecker product of two matrices $B$, $C$ of arbitrary size. |
| $*$ | Convolution operator, i.e., $(g * u)_k = \sum_{l=0}^{\infty} g_l u_{k-l}$. |
| $\nabla \rho(\bar{x})$ | For a continuously differentiable function $\rho : \mathbb{R}^n \to \mathbb{R}$, $\nabla \rho(\bar{x})$ denotes the gradient vector field of $\rho$ w.r.t. $x$ evaluated at the point $\bar{x} \in \mathbb{R}^n$. |
| $\mathtt{H}_\rho(\bar{x})$ | For a twice continuously differentiable function $\rho : \mathbb{R}^n \to \mathbb{R}$, $\mathtt{H}_\rho(\bar{x}) \in \mathbb{R}^{n \times n}$ denotes the matrix of second order partial derivatives of $\rho$ w.r.t. $x$ evaluated at the point $\bar{x} \in \mathbb{R}^n$ (Hessian matrix). |
| $T_u \mathbb{S}^{n-1}$ | Tangent space of $\mathbb{S}^{n-1}$ at $u \in \mathbb{S}^{n-1}$. |
| $G^\sim(z)$ | Para-hermitian conjugate of a complex matrix-valued function $G(z)$, i.e., $G^\sim(z) = G^\top(z^{-1})$. |

# Abstract

Due to their relevance in systems analysis and controller design, we consider the problem of determining system-theoretic input-output properties of linear time-invariant (LTI) systems. While, in practice, the input-output behavior of dynamic systems is often undisclosed and obtaining a suitable mathematical model via first principles can be a cumbersome task, data of the system in form of input-output trajectories is often and increasingly available. Therefore, we present different methods to determine system-theoretic input-output properties directly from data without deriving or identifying a mathematical model first. In particular, we study iterative methods, where data is actively sampled by performing experiments on the unknown system, as well as approaches based on available (offline) data. Considering these offline approaches, we first develop necessary and sufficient conditions requiring only one input-output trajectory to certify system properties on the basis of Willems' fundamental lemma and then introduce robust approaches providing guaranteed bounds on the respective input-output property in the case of noisy data.

**Iterative methods.** We generalize state-of-the-art iterative sampling schemes determining the $\mathcal{L}_2$-gain of an LTI system to a more general framework in Chapter 3, which includes the consideration of passivity properties and conic relations. Such iterative schemes require active sampling of input-output data by performing experiments or simulations on the unknown system with iteratively updated input signals. We first show how the respective input-output system properties can be reformulated in terms of optimization problems and present sampling strategies based on gradient dynamical systems and saddle point flows to find their optimizers. These sampling strategies are based on the fact that the gradients of the optimization problems can be evaluated from only input-output data samples even though the input-output behavior of the system is unknown. This leads us to evolution equations, whose convergence properties are then discussed in continuous time and discrete time. Multiple numerical examples show the potential and applicability of the introduced approaches.

**Offline methods.** Since the requirement of iterative experiments can be limiting, we introduce a necessary and sufficient condition for LTI systems to verify input-output system properties from only one input-output trajectory on the basis of Willems' fundamental lemma in Chapter 4. More specifically, we consider the general classes of dissipativity properties and integral quadratic constraints (IQCs). For important classes of such input-output system properties, we provide convex optimization problems in form of semidefinite programs (SDPs) to retrieve the optimal, i.e., the tightest, system property description that is satisfied by the unknown system. Furthermore, we provide insights and results on the difference between finite and infinite horizon IQCs and finally illustrate the effectiveness of the proposed scheme in a variety of simulation studies including noisy measurements and a high dimensional system.

**Robust methods.** While the previous iterative and offline approaches for data-driven dissipativity analysis guarantee the dissipativity condition only over a finite-time horizon and provide no quantitative guarantees on robustness in the presence of noise, we provide a framework to verify dissipativity properties from noisy data with desirable guarantees in Chapter 5. We first consider the case of input-state measurements, where we provide nonconservative and computationally attractive dissipativity conditions in the presence of unknown but bounded process noise. We then extend this approach to input-output data in the noise-free as well as in the noisy case. Finally, we apply the proposed approach to real-world data of a two-tank water system and illustrate its applicability and advantages compared to established methods based on system identification.

The main goal of this thesis is to introduce a framework to determine input-output system properties directly from data without deriving or identifying a mathematical model first. To this end, we develop different methods based on various control-theoretic results and insights, which all have their individual advantages, limitations, and their corresponding application cases.

# Deutsche Kurzfassung

Aufgrund ihrer Relevanz für Systemanalyse und Reglerentwurfsverfahren entwickeln wir Methoden zur Bestimmung von systemtheoretischen Eingangs-/Ausgangseigenschaften linearer, zeitinvarianter Systeme. Während in der Praxis das genaue Eingangs-/Ausgangsverhalten solcher Systeme oft unbekannt ist und das Aufstellen einer geeigneten mathematischen Beschreibung über physikalische Modellierung eine mühsame Aufgabe sein kann, sind Daten in Form von Eingangs-/Ausgangstrajektorien oft und in zunehmendem Maße verfügbar. Deshalb präsentieren wir unterschiedliche Methoden zur Bestimmung systemtheoretischer Eingangs-/Ausgangseigenschaften direkt aus Daten, für deren Anwendung nicht zuerst ein mathematisches Modell aufgestellt oder identifiziert werden muss. Im Speziellen untersuchen wir iterative Methoden, bei denen wiederholt Experimente am unbekannten System mit iterativ angepasstem Eingangssignal durchgeführt werden, sowie weitere Herangehensweisen, die auf bereits verfügbaren ('offline') Daten basieren. In der Kategorie der offline Methoden entwickeln wir zum einen Bedingungen für Systemeigenschaften basierend auf Willems' fundamentalem Lemma, welche nur eine Eingangs-/Ausgangstrajektorie benötigen, und zum anderen robuste Methoden, die auch im Fall von Störungen oder Rauschen in den Daten Systemeigenschaften garantieren können.

**Iterative Methoden.** Wir generalisieren bisher übliche iterative Verfahren zur Bestimmung der $\mathcal{L}_2$-Verstärkung von linearen, zeitinvarianten Systemen zu einem allgemeineren Analyseverfahren, welches auch weitere Systemeigenschaften wie Passivität und Kegelrelationen (englisch: *conic relations*) untersuchen kann. Solche iterative Verfahren basieren auf wiederholten Experimenten am unbekannten System mit iterativ angepassten Eingangssignalen. Die iterativen Verfahren zur Bestimmung von Eingangs-/Ausgangseigenschaften, die wir in Kapitel 3 vorstellen, basieren auf dynamischen Gradientensystemen und Sattelpunktflüssen, wobei die Gradienten aus Eingangs-/Ausgangsdaten ausgewertet werden können. Das führt uns zu Evolutionsgleichungen, deren Konvergenzeigenschaften wir in kontinuierlicher und diskreter Zeit diskutieren. Mehrere numerische Beispiele zeigen das Potenzial und die Anwendbarkeit der eingeführten Methoden.

**Offline Methoden.** Da die Erfordernis von iterativen Experimenten für bestimmte Anwendungen einschränkend sein kann, stellen wir in Kapitel 4 eine notwendige und hinreichende Bedingung zur Zertifizierung von Eingangs-/Ausgangseigenschaften linearer zeit-invarianter Systeme aus nur einer Eingangs-/Ausgangstrajektorie vor. Im Speziellen entwickeln wir datenbasierte Methoden, die basierend auf Willems' fundamentalem Lemma allgemeine Dissipativitätseigenschaften und integral-quadratische Beschränkungen (englisch: *integral quadratic constraints*, IQCs) verifizieren können. Für wichtige Klassen solcher Eingangs-/Ausgangseigenschaften entwickeln wir konvexe Optimierungsprobleme in Form von semidefiniten Programmen (englisch: *semidefinite programs*, SDPs), welche die optimale, das heißt die restriktivste, Beschreibung der Systemeigenschaft bestimmen, die das unbekannte System erfüllt. Zusätzlich untersuchen wir den Unterschied zwischen der Betrachtung einer IQC über den endlichen und den unendlichen Zeithorizont. Abschließend veranschaulichen wir die Anwendbarkeit des vorgeschlagenen Ansatzes anhand verschiedener Simulationsstudien, die unter anderem verrauschte Messungen und ein hochdimensionales Beispiel beinhalten.

**Robuste Methoden.** Die vorgestellten iterativen und offline Methoden für datenbasierte Dissipativitätsanalyse garantieren die Dissipativitätsbedingung nur über einen endlichen Zeithorizont und liefern darüber hinaus keine quantitativen Garantien bezüglich Robustheit gegenüber Rauschen oder Störungen in den Daten. Deshalb erweitern wir in Kapitel 5 unsere Betrachtungen auf die Zertifizierung von Dissipativitätseigenschaften aus verrauschten Daten mit den angestrebten Garantien. Zuerst betrachten wir den Fall von Eingangs-/Zustandsmessungen, für den wir eine Methode zur Berechnung garantierter Systemeigenschaften aus Daten mit unbekanntem, aber beschränktem Prozessrauschen entwickeln, die weder konservativ noch rechenintensiv ist. Diesen Ansatz erweitern wir dann auf den Fall rauschfreier und rauschbehafteter Eingangs-/Ausgangsdaten. Zuletzt wenden wir die vorgestellte Herangehensweise auf reale Daten eines Mehrtanksystems an und veranschaulichen ihre Anwendbarkeit und ihre Vorteile gegenüber etablierten, auf Systemidentifikation basierenden Ansätzen.

Das primäre Ziel dieser Doktorarbeit ist die Einführung eines umfassenden Konzepts für die Verifizierung von Eingangs-/Ausgangseigenschaften direkt aus Daten, ohne zuerst ein mathematisches Modell aufzustellen oder zu identifizieren. Dafür entwickeln wir basierend auf unterschiedlichen regelungstechnischen Resultaten und Erkenntnissen verschiedene Methoden, die jeweils individuelle Vorteile, Einschränkungen sowie Anwendungsfälle haben.

# Chapter 1

# Introduction

## 1.1 Motivation

Most state-of-the-art systems analysis and controller design methods are based on an accurate mathematical model that adequately describes the system at hand. While the available literature on model-based systems analysis and controller design is quite elaborate and comes with rigorous guarantees for stability, performance, and robustness, acquiring a suitable and accurate mathematical model of the plant via first principles can be a time-consuming task that depends on expert knowledge. At the same time, data are becoming ubiquitous, cheap, and increasingly available in engineering applications in form of input-output trajectories of dynamic systems from experiments or simulations. These data potentially capture sufficient information about the otherwise unknown system needed for systems analysis and controller design. Therefore, there has recently been an increasing interest in the field of data-driven methods for systems analysis and control that systematically extract and directly exploit the information captured in the data.

This increasing interest in data-driven methods led to a significantly growing area of data-driven control approaches. In the survey paper [40] on data-driven controller design techniques, some rather established approaches to data-driven control are summarized such as virtual reference feedback tuning [17], iterative feedback tuning [39], and unfalsified control [45]. Since the appearance of this survey in 2013, many more approaches in the field of data-driven control have emerged and the literature is expanding rapidly. These recent approaches include stabilizing, optimal or robust control for linear time-invariant (LTI) systems on the basis of data-driven closed-loop parametrizations [75, 102, AK3], data-driven predictive control exploiting Willems' fundamental lemma [8, 20, 111], linear quadratic regulator (LQR) tuning via Gaussian process optimization [50, 66], data-driven control of networks and multi-agent settings [4, 24, AK24], mentioning only a few examples.

With the increasing availability of and interest in data, also the well-established field of system identification or data-driven modeling has seen a renewed interest. While the general area of system identification has a long history and provides a well-established and extensive amount of literature (see, e.g., [47] and references therein), there are still open questions. Even for LTI systems, finding nonasymptotic guarantees on a model from noisy data is yet a challenging problem [97] and part of ongoing research. Some promising results in this direction have been presented in [56], where the authors identify a model with guaranteed bounds on the uncertainty from data that are perturbed by stochastic noise, which can then be accounted for in robust controller design approaches [21]. Two further results in this direction include [74, 91] providing nonasymptotic system identification guarantees. However, all of these approaches require rather strong assumptions on the data and the noise. In [21], for example, all but the last state transition in each trajectory are discarded, in [91] state measurements corrupted by Gaussian independent and identically distributed (i.i.d.) noise are required, and assumptions in [74] include Gaussian i.i.d. noise and input, known system order $n$, and zero initial condition. System identification approaches in the case of deterministic noise typically rely on set membership estimation, where the key challenge is in providing both nonconservative but computationally tractable error bounds [61].

One complementary approach to the direct controller design from data or to identifying a full mathematical model of the system, respectively, is to learn and analyze certain system-theoretic input-output properties from data. Such input-output properties not only provide valuable insights into the system, but can also be leveraged to design a controller. In fact, properties such as dissipativity properties or integral quadratic constraints (IQCs) allow for the direct application of well-known feedback theorems for controller design, as shown for example in [22, 23, 99, 113]. These controllers come with desirable guarantees on stability and robustness, which is still an open problem for many data-driven control approaches. Furthermore, by first determining input-output properties, the controller structure is not a priori determined or parametrized in contrast to many data-driven control and data-driven tuning approaches. At the same time, learning such system properties from data can retain many of the desired advantages of data-driven methods compared to system identification approaches. They can be simple to apply without requiring expert knowledge, avoid the computational load of identifying a full mathematical model, and skip a possibly unnatural fit to a parametric system model potentially introducing additional error. One specifically valuable application area of input-output system properties includes cooperative control and multi-agent settings, where dissipativity can be used for compositional certification of stability, performance, and safety [2]. Besides the application of controller design via well-known feedback theorems, applications of input-output system properties from data also include controller validation [36] as well as fault detection and mitigation [112].

Therefore, we aim to provide a framework to determine system-theoretic input-output properties such as dissipativity properties and IQCs directly from data with guarantees. We strive to introduce methods that are easy to apply, computationally attractive and do not require expert knowledge. In this thesis, we focus on approaches to determine input-output properties of LTI systems, while we will discuss potential extensions to nonlinear systems in the individual summaries as well as in the conclusions of this thesis.

## 1.2 Related work

Due to the relevance of input-output properties in systems analysis and control and the well-established literature on dissipativity-based controller design, there has been a considerable number of approaches to determine such properties from data. Thus, the following literature review summarizes existing results for determining input-output properties from data as well as points out related methods and approaches.

Very generally, the literature on data-driven verification of input-output system properties can be roughly categorized into three conceptually different setups, as presented in the following.

#### Input-output properties of nonlinear systems from big data

Firstly, there exist some interesting ideas for determining input-output system properties for rather general classes of nonlinear systems from large amounts of input-output data tuples that are stored and available for analysis. In [62], the authors derive overestimates on the $\mathcal{L}_2$-gain, the shortage of passivity, and a cone containing all input-output samples based on finite, but densely sampled, input-output data. This idea was further analyzed in [89] and extended to nonlinearity measures in [54]. All these methods are based on Lipschitz assumptions on the nonlinear system, exploited in a similar fashion also in nonlinear data-driven control approaches [68, 69]. More general dissipation inequalities were considered in [AK20], where the ordering of the supply rates via the S-procedure allows to infer system properties from only a finite amount of input-output data. Another approach to determine passivity properties via Gaussian process optimization was introduced in [AK23], which additionally allows for active sampling schemes (i.e., choosing an input and performing an experiment) to increase data efficiency. This idea was extended to a deterministic setting in [53] via successive graph approximation to infer nonlinearity measures.

However, to receive quantitative bounds on certain dissipation inequalities, all these approaches require immense amounts of input-output trajectories, which limits their applicability. While we focus only on LTI systems in this thesis, we discuss extensions for certain classes of nonlinear systems, which potentially reduce the conservatism or the required number of data samples significantly compared to the existing literature on data-driven analysis of nonlinear systems introduced above.

#### Input-output properties of LTI systems from iterative sampling schemes

The second category of data-driven systems analysis methods is based on input signal optimization and iterative experiments for LTI systems, closely related to the general idea of iterative learning control (ILC), see, e.g., [14, 67]. The basic idea for iterative data-driven systems analysis has already been introduced in 2005 [38, Section 12.2], where the author applies a so-called power iteration method to receive an $\mathcal{L}_2$-gain estimate of the otherwise unknown system. The power iteration method is based on the well-known power method to determine the dominant eigenvector and eigenvalue pair of a matrix (see, e.g., [31]). It can be shown that this power iteration method, based on iterative experiments, is guaranteed to converge to the gain of the system [106], which has been further analyzed in terms of asymptotic statistical properties in [105]. While the method proposed in [106] requires two experiments per iteration, it is shown in [82] how this can be reduced to one experiment per iteration. In [72], this sampling scheme is extended to multiple-input multiple-output (MIMO) systems and applied to robust active vibration isolation.

Another similar approach is related to game theory, where the input domain is discretized at specific frequencies and the inputs over the iterations are adapted such that the excitation level is increased in those regions of the frequency domain where the peak is expected to be located [79]. Along the same lines, the multi-armed bandit approach is adopted to $\mathcal{L}_2$-gain estimation in [64] with the goal to maximize the probability of choosing the arm drawing samples, i.e., frequencies, with the highest amplitude gain in the output. This approach has been extended in [63] and improved in [65] by combining power iterations and Weighted Thompson Sampling.

However, all these iterative approaches for LTI systems [63, 64, 65, 72, 79, 82, 105, 106] only consider estimating the $\mathcal{L}_2$-gain. In Chapter 3 in this thesis, we generalize the idea of input signal optimization and iterative experiments for LTI systems to infer the $\mathcal{L}_2$-gain to a more general framework, which includes the consideration of passivity properties and conic relations.

#### Input-output properties of LTI systems from one offline trajectory

Finally, the third category of methods to verify dissipativity properties from data includes offline computational approaches from one input-state or input-output trajectory of LTI systems. In this direction, Willems' fundamental lemma [108] offers a fruitful basis by proving that the full behavior of an LTI system can be described by a Hankel matrix containing one previously measured input-output trajectory, given that the input component is persistently exciting. This result, as discussed in [9, 104], provides an equivalent description of LTI systems based only on data, which hence allows for systems analysis and control with desirable guarantees from input-output trajectories. Contributions to data-driven approaches based on Willems' fundamental lemma range from data-driven simulation and output-matching [52] to data-driven controller design [109, AK27] and data-driven model predictive control [8, 20, 111].

Considering data-driven systems analysis, Willems' fundamental lemma allows to optimize over the input signals not via iterative experiments but offline from only one input-output trajectory since it provides a data-based characterization of all trajectories of an unknown LTI system. The idea of determining dissipativity from a one-shot trajectory on the basis of the behavioral framework via Willems' fundamental lemma was introduced in [57]. However, their approach results in a nonconvex indefinite quadratic program, which is generally very hard to solve. Therefore, we introduce in Chapter 4 of this thesis how dissipativity properties, and more generally IQCs, can be certified by verifying positive semidefiniteness of a single data-dependent matrix.

One limitation to all of the introduced data-driven dissipativity results is, however, that no quantitative guarantees can be provided in the case of noisy data. To approach this, we make use of another fruitful line of work related to Willems' fundamental lemma, which is based on the idea of finding a parameterization of the closed loop from input-state trajectories of an otherwise unknown system [75]. This idea becomes especially interesting when it comes to noise-corrupted trajectories, where a suitable parameterization can represent all possible closed loops that are consistent with the data [AK3], hence allowing for robust data-driven state-feedback design from noise-corrupted data. This idea has been further extended and improved, e.g., in [10, 102]. In Chapter 5 of this thesis, we utilize this line of work to derive rigorous deterministic bounds on dissipativity properties even from noise-corrupted data.

## 1.3 Contribution and outline of this thesis

In the following, we detail the outline of this thesis and clarify the contributions.

#### Chapter 2: Background

In Chapter 2, we present the basic problem setup underlying this thesis. In particular, we define and discuss the input-output system properties of interest. Furthermore, we provide equivalent representations from different viewpoints of such system properties, which will be exploited throughout the thesis.

#### Chapter 3: Iterative schemes to determine input-output properties

The thrust of Chapter 3 is to present a systematic approach to iteratively determine certain dissipativity properties from input-output samples, where the input-output map remains undisclosed. To this end, we introduce different sampling strategies that can be summarized by (i) formulating the system property of interest as an optimization problem, and (ii) iteratively performing experiments that update the input along the gradient of the resulting optimization problem. In particular, we use multiple input-output trajectories from iterative experiments to investigate the $\mathcal{L}_2$-gain, the shortage of passivity and conic relations. We start in Section 3.1 with discrete-time LTI systems with a thorough analysis of continuous-time optimization as well as the implications for the iterative scheme (discrete-time optimization), where advanced sampling schemes can improve the convergence rate. In Section 3.2, we generalize the framework presented in Section 3.1 by, firstly, showing how the introduced iterative approach to determine input-output system properties can be extended to continuous-time LTI systems in Section 3.2.1. Secondly, we extend the framework to MIMO systems in Section 3.2.2 and additionally provide results on the robustness of the presented framework to measurement noise in Section 3.2.3. Finally, we apply the introduced approaches to different simulation examples in Section 3.3, including an oscillating system and a high dimensional system, and we end with a short summary in Section 3.4.

The results of Chapter 3 have been previously presented in [AK8, AK19, AK21, AK22].

**Chapter 4: Offline approaches to determine input-output properties**

In Chapter 4, we provide a computationally simple approach to certify input-output system properties from offline data. More precisely, we introduce necessary and sufficient conditions for a discrete-time LTI system to satisfy dissipativity properties, or more generally IQCs, given only one input-output data trajectory in Section 4.1. These conditions can be verified by simply checking whether a single data-dependent matrix is positive semidefinite, with the only requirements being i) a persistently exciting (but otherwise arbitrary) input signal and ii) knowledge of an upper bound on the lag of the system. This theory is based on Willems' fundamental lemma [108], a result originally developed in the context of behavioral systems theory, which provides a data-based characterization of all trajectories of an unknown LTI system. On the basis of the developed simple condition to infer IQCs from data, we extend the approach to noisy measurements, where we provide a very promising heuristic relaxation. Additionally, we characterize optimal system properties and provide semidefinite programs (SDPs) to find such IQCs in Section 4.2, which can provide more informative and tighter descriptions of the unknown system allowing for, e.g., less conservative robust controller designs. While in most of Chapter 4 the IQC property is only considered over a finite-time horizon, we infer bounds on the respective system property over the infinite-time horizon in Section 4.3. We conclude this chapter with simulation studies on a high dimensional numerical example in Section 4.4, demonstrating the potential of the introduced approach, together with a short summary in Section 4.5.

The results of Chapter 4 have been previously presented in [AK4, AK16].

### Chapter 5: Bounds on input-output properties via a robust viewpoint

In Chapter 5, we develop guarantees on dissipativity properties from trajectories corrupted by bounded process noise. To this end, we first introduce an equivalent dissipativity characterization on the basis of one input-state trajectory in Section 5.1. In contrast to many other approaches to determine input-output properties from data (e.g., iterative schemes [106, AK8] and methods based on Willems' fundamental lemma [57, AK16]), we exploit the state-space definition of dissipativity in this chapter, which can be verified by taking a difference viewpoint, i.e., looking at the difference at two time points. This yields the advantage that guarantees on system properties over the infinite horizon can be obtained. More importantly, this allows to introduce a rigorous treatment of noisy data in Section 5.2, where we provide a computationally attractive and nonconservative robust verification framework for dissipativity properties from input-state data corrupted by bounded process noise. To this end, we first characterize all system matrices that are consistent with the data and an a priori known noise bound and then verify the dissipativity property of interest for all systems in this set. As the requirement of state measurements can be restrictive in practice, we extend the results to input-output trajectories in the noise-free case in Section 5.3, followed by a consideration of noise-corrupted input-output trajectories in Section 5.4. This finally yields a method to guarantee dissipativity properties over the infinite horizon from one noise-corrupted input-output trajectory of finite length via SDPs. Lastly, we apply the introduced approach to real-world data of a two-tank water system in Section 5.5 and conclude with a short summary in Section 5.6.

The results of Chapter 5 have been previously presented in [AK5, AK6].

### Chapter 6: Conclusions

In this chapter, we summarize the contributions of this thesis, contrast the different approaches to data-driven inference of input-output system properties, and provide some directions for future research.

# Chapter 2

# Background

In this chapter, we present the general problem setup considered in this thesis and introduce the notions of dissipativity as well as IQCs.

## 2.1 Problem setup

In this thesis, we consider discrete-time LTI systems

$$\begin{aligned} x_{k+1} &= Ax_k + Bu_k, \\ y_k &= Cx_k + Du_k, \end{aligned} \tag{2.1}$$

with $x_k \in \mathbb{R}^n$, $u_k \in \mathbb{R}^m$, and $y_k \in \mathbb{R}^p$, $k \in \mathbb{N}$, where $(A, B, C, D)$ define a minimal realization.

A characteristic of such a controllable and observable LTI system, which will be of special importance in Chapters 4 and 5 of this thesis, is the lag $\underline{l}$ of the system.

**Definition 2.1.** *The lag $\underline{l}$ of System (2.1) is the smallest integer $l \in \mathbb{N}_+$ such that the observability matrix given by*

$$O_l = \begin{bmatrix} C \\ CA \\ \vdots \\ CA^{l-1} \end{bmatrix}$$

*has rank $n$.*

Note that, while we restrict our attention to the analysis of systems given in a minimal realization, the approaches introduced in this thesis naturally extend to systems which have the same input-output behavior as a minimal realization. A more detailed explanation can be found in Chapter 5, Section 5.3 and Section 5.4, where this extension will be of special importance.

We are interested in the case where no model of the given system is known, but data in form of input-output trajectories of the system are available.

**Definition 2.2.** *We say that an input-output sequence* $\{u_k, y_k\}_{k=0}^{N-1}$ *is a trajectory of System* (2.1) *if there exists an initial condition* $\bar{x} \in \mathbb{R}^n$ *such that*

$$\begin{aligned} x_{k+1} &= Ax_k + Bu_k, \;\; x_0 = \bar{x}, \\ y_k &= Cx_k + Du_k, \end{aligned}$$

*for* $k = 0, \ldots, N-1$.

On the basis of such input-output trajectories, we aim to determine input-output properties of the unknown system.

One measure that captures the informativity of a given input-output trajectory is the notion of persistency of excitation, which is of central importance in the field of system identification, adaptive control, and data-driven approaches, see, e.g., [15, 51, 108]. To recall the notion of persistency of excitation, we define the Hankel matrix of a finite sequence $\{u_k\}_{k=0}^{N-1}$ by

$$H_L(u) = \begin{bmatrix} u_0 & u_1 & \ldots & u_{N-L} \\ u_1 & u_2 & \ldots & u_{N-L+1} \\ \vdots & \vdots & \ddots & \vdots \\ u_{L-1} & u_L & \ldots & u_{N-1} \end{bmatrix}. \tag{2.2}$$

**Definition 2.3.** *We say that a sequence* $\{u_k\}_{k=0}^{N-1}$ *with* $u_k \in \mathbb{R}^m$ *is persistently exciting of order* $L$ *if* $\operatorname{rank}(H_L(u)) = mL$.

This definition is of special importance to characterize the informativity of the input-output trajectory in the case of given (offline) data, as presented in Chapter 4 and Chapter 5. Note that Definition 2.3 implies $N \geq (m+1)L - 1$, yielding a lower bound on the length of the given input signal for being persistently exciting of order $L$, which also depends on the number of inputs $m$.

## 2.2 Dissipativity properties

Since their introduction in [107], dissipativity properties have become increasingly relevant in systems analysis and control. Usually, these properties can be verified using a full mathematical model of the system. In this thesis, we are interested in determining input-output properties, and in particular dissipativity properties, directly from data without first identifying an explicit model. While the notion of dissipativity was introduced in [107] for general (nonlinear) systems and supply rates, we make use of equivalent formulations for LTI systems with quadratic supply rates as, e.g., presented in [87].

Quadratic supply rates are functions $s : \mathbb{R}^m \times \mathbb{R}^p \to \mathbb{R}$ defined by

$$s(u_k, y_k) = \begin{bmatrix} u_k \\ y_k \end{bmatrix}^\top \Pi \begin{bmatrix} u_k \\ y_k \end{bmatrix}, \tag{2.3}$$

where we often use a partitioning of the matrix $\Pi \in \mathbb{R}^{(m+p)\times(m+p)}$ given by

$$\Pi = \begin{bmatrix} R & S^\top \\ S & Q \end{bmatrix}$$

with $Q = Q^\top \in \mathbb{R}^{p\times p}$, $S \in \mathbb{R}^{m\times p}$ and $R = R^\top \in \mathbb{R}^{m\times m}$.

**Definition 2.4.** *We say that System (2.1) is dissipative w.r.t. the supply rate $s$ if there exists a storage function $V : \mathbb{R}^n \to \mathbb{R}$ which is bounded from below such that*

$$V(x_{k''}) - V(x_{k'}) \leq \sum_{k=k'}^{k''-1} s(u_k, y_k) \tag{2.4}$$

*holds for all $0 \leq k' < k''$ and all signals $(u, x, y)$ which satisfy (2.1). It is said to be strictly dissipative if instead of (2.4)*

$$V(x_{k''}) - V(x_{k'}) \leq \sum_{k=k'}^{k''-1} s(u_k, y_k) - \epsilon \sum_{k=k'}^{k''-1} \|u_k\|^2$$

*holds for all $0 \leq k' < k''$, all signals $(u, x, y)$ which satisfy (2.1), and some $\epsilon > 0$.*

Hereby, the matrices $(Q, S, R)$ in the supply rate define the specific system property at hand. With the supply rates defined by

$$\begin{aligned} \Pi_\gamma &= \begin{bmatrix} \gamma^2 I & 0 \\ 0 & -I \end{bmatrix}, \quad \Pi_\nu = \begin{bmatrix} -\nu I & 0.5I \\ 0.5I & 0 \end{bmatrix}, \\ \Pi_\beta &= \begin{bmatrix} 0 & 0.5I \\ 0.5I & \beta I \end{bmatrix}, \quad \Pi_c = \begin{bmatrix} r^2 I - C^\top C & C^\top \\ C & -I \end{bmatrix}, \end{aligned} \tag{2.5}$$

to name the most prominent examples, we retrieve the $\mathcal{L}_2$-gain (or operator gain) $\gamma$, the input feedforward passivity parameter $\nu$, the shortage of passivity $\beta$, and a conic relation with center matrix $C$ and radius $r$, respectively. These specific dissipativity properties will be the main focus of Chapter 3, whereas we address more general input-output properties in Section 4 and Section 5. The general dissipativity property specified by $(Q, S, R)$ will in the following also be referred to as $(Q, S, R)$-dissipativity.

In this thesis, we make use of different equivalent conditions on dissipativity of an LTI system. The following standard result together with explanations and the proof can be found, e.g., in [87] and references therein.

**Theorem 2.1.** *Let $s$ be a quadratic supply rate of the form (2.3). Then the following statements are equivalent.*

*a) System (2.1) is $(Q, S, R)$-dissipative.*

*b) There exists a quadratic storage function $V(x) = x^\top \mathcal{X} x$ with $\mathcal{X} = \mathcal{X}^\top \succeq 0$ such that*

$$V(x_{k+1}) - V(x_k) \leq s(u_k, y_k)$$

*holds for all $k$ and all $(u, x, y)$ satisfying (2.1).*

*c) There exists a matrix $\mathcal{X} = \mathcal{X}^\top \succeq 0$ such that*

$$\begin{bmatrix} A^\top \mathcal{X} A - \mathcal{X} - \hat{Q} & A^\top \mathcal{X} B - \hat{S} \\ (A^\top \mathcal{X} B - \hat{S})^\top & -\hat{R} + B^\top \mathcal{X} B \end{bmatrix} \preceq 0 \tag{2.6}$$

*with $\hat{Q} = C^\top Q C$, $\hat{S} = C^\top S + C^\top Q D$, and $\hat{R} = D^\top Q D + (D^\top S + S^\top D) + R$.*

Another equivalent standard condition for dissipativity is the input-output formulation of dissipativity, for which further explanations and the proof can be found in [37].

**Theorem 2.2.** *System (2.1) is dissipative w.r.t. the supply rate s in (2.3) according to Definition 2.4 if and only if*

$$\sum_{k=0}^{h} s(u_k, y_k) \geq 0, \quad \forall h \geq 0, \tag{2.7}$$

*for all trajectories* $\{u_k, y_k\}_{k=0}^{\infty}$ *of System (2.1) with* $u \in l_2^m$ *and initial condition* $x_0 = 0$.

In much of the existing literature on input-output system properties from data, as well as in most parts of Chapter 3 and Chapter 4, the relaxed version of finite-horizon dissipativity, or $L$-dissipativity, is considered, which was introduced in [57].

**Definition 2.5.** *We say that System (2.1) is L-dissipative w.r.t. the supply rate s in (2.3) if*

$$\sum_{k=0}^{h} s(u_k, y_k) \geq 0, \quad \forall h = 0, \ldots, L-1, \tag{2.8}$$

*for all trajectories* $\{u_k, y_k\}_{k=0}^{L-1}$ *of System (2.1) with initial condition* $x_0 = 0$.

However, for LTI systems it is actually sufficient for $L$-dissipativity that the inequality in (2.8) only holds over the horizon $L$, as shown in the following proposition.

**Proposition 2.1.** *System (2.1) is L-dissipative w.r.t. the supply rate s in (2.3) if*

$$\sum_{k=0}^{L-1} s(u_k, y_k) \geq 0 \tag{2.9}$$

*holds for all trajectories* $\{u_k, y_k\}_{k=0}^{L-1}$ *of System (2.1) with initial condition* $x_0 = 0$.

*Proof.* We show that the stated condition for a time horizon $L = L_1$ implies the same condition for any shorter time horizon $L_2 \leq L_1$.

Since (2.9) must hold for all inputs $\{u_k\}_{k=0}^{L_1-1}$, it must also hold for all $\{u_k\}_{k=0}^{L_1-1} \in \mathcal{U}$ with $\mathcal{U} = \{\{u_k\}_{k=0}^{L_1-1} | u_k = 0,\ k = 0, \ldots, L_1 - L_2 - 1\}$. For all $\{u_k\}_{k=0}^{L_1-1} \in \mathcal{U}$, moreover, $x_{L_1-L_2} = 0$ since the system is LTI and $x_0 = 0$. The transformation $\bar{k} = k - (L_1 - L_2)$ then shows that (2.9) must also hold for $L = L_2$ for all $\{u_k\}_{k=0}^{L_2-1}$, which proves the claim. ■

The above proposition also shows that dissipativity implies $L$-dissipativity for any $L$. On the other hand, if $L$-dissipativity holds for arbitrarily large $L$ (i.e., taking the limit $L \to \infty$), then this in turn implies dissipativity. A more detailed discussion on this topic and a thorough analysis on the case of large but finite choices of $L$ can be found in Section 4.3.

In the remainder of this thesis, we use the equivalences stated in Theorem 2.1 and Theorem 2.2 to verify or find dissipativity properties from data.

## 2.3 Integral quadratic constraints

Another more general class of input-output system properties are IQCs as introduced in [58]. To introduce the notion of IQCs in discrete time, we follow the presentation and notation in [27, 41, 60]. Let $P \in \mathcal{RL}_\infty^{(m+p)\times(m+p)}$ be a linear, bounded, self-adjoint operator. Then a system is said to satisfy the IQC defined by the multiplier $P$ if

$$\frac{1}{2\pi}\int_{-\pi}^{\pi} \begin{bmatrix} \hat{u}(e^{i\omega}) \\ \hat{y}(e^{i\omega}) \end{bmatrix}^* P(e^{i\omega}) \begin{bmatrix} \hat{u}(e^{i\omega}) \\ \hat{y}(e^{i\omega}) \end{bmatrix} d\omega \geq 0 \tag{2.10}$$

holds for all $u \in l_2^m$, where $\hat{u}$ and $\hat{y}$ are the discrete-time Fourier transforms of the input $u$ and the corresponding output $y$, respectively, and the superscript $^*$ denotes complex conjugation.

By Parseval's Theorem, the IQC in (2.10) can also be evaluated in the time domain [60], which brings us to the time-domain IQC formulation (cf. [41]) based on a factorization $P(z) = \Psi^\sim(z)M\Psi(z)$ where $M = M^\top \in \mathbb{R}^{n_r \times n_r}$ and $\Psi \in \mathcal{RH}_\infty^{n_r\times(m+p)}$. A factorization of such a form $P(z) = \Psi^\sim(z)M\Psi(z)$ is always possible, although it is in general not unique. In fact, one can always construct a factorization from any $P$ such that $\Psi$ is causal and stable [41]. Hence, we consider $\Psi$ to be a stable and causal LTI system with zero initial condition, which we will also refer to as a filter $\Psi$. This finally leads us to the following definition of IQCs in the discrete-time domain [27].

**Definition 2.6.** *We say that System (2.1) satisfies an IQC for a given multiplier $P(z) = \Psi^\sim(z)M\Psi(z)$ if*

$$\sum_{k=0}^{h} r_k^\top M r_k \geq 0, \quad \forall h \geq 0, \quad \textit{with} \quad r_k = \left(g^\Psi * \begin{bmatrix} u \\ y \end{bmatrix}\right)_k, \tag{2.11}$$

*for all trajectories $\{u_k, y_k\}_{k=0}^\infty$ of System (2.1) with $u \in l_2^m$ and initial condition $x_0 = 0$, where $\{g_k^\Psi\}_{k=0,1,2,\ldots}$ denotes the impulse response of $\Psi$ with $g_k^\Psi \in \mathbb{R}^{n_r\times(m+p)}$.*

Intuitively, computing the sequence $\{r_k\}_{k=0}^{\infty}$ corresponds to filtering the input and output signals $\{u_k, y_k\}_{k=0}^{\infty}$ through an LTI system $\Psi$ with zero initial condition (here simply denoted by the convolution with the impulse response of $\Psi$). The time-domain IQC is then an inequality ('dissipativity condition') on the filtered output $\{r_k\}_{k=0}^{\infty}$.

Analogously to $L$-dissipativity in Definition 2.5, we introduce the relaxed version of finite-time IQCs, or in short $L$-IQCs. A discussion and results on the connection of IQCs and $L$-IQCs can be found in Section 4.3.

**Definition 2.7.** *We say that System* (2.1) *satisfies an L-IQC for a given multiplier* $P(z) = \Psi^{\sim}(z)M\Psi(z)$ *if*

$$\sum_{k=0}^{h} r_k^\top M r_k \geq 0, \quad \forall h = 0, \ldots, L-1, \quad \text{with} \quad r_k = \left( g^{\Psi} * \begin{bmatrix} u \\ y \end{bmatrix} \right)_k, \tag{2.12}$$

*holds for all trajectories* $\{u_k, y_k\}_{k=0}^{L-1}$ *of System* (2.1) *with initial condition* $x_0 = 0$, *where* $\{g_k^{\Psi}\}_{k=0,1,2,\ldots}$ *denotes the impulse response of* $\Psi$ *with* $g_k^{\Psi} \in \mathbb{R}^{n_r \times (m+p)}$.

As in the case of $L$-dissipativity, it is sufficient for an $L$-IQC to verify that (2.12) holds over the horizon $L$.

**Proposition 2.2.** *System* (2.1) *satisfies an L-IQC for a given multiplier* $P(z) = \Psi^{\sim}(z)M\Psi(z)$ *if and only if*

$$\sum_{k=0}^{L-1} r_k^\top M r_k \geq 0, \quad \text{with } r_k = \left( g^{\Psi} * \begin{bmatrix} u \\ y \end{bmatrix} \right)_k, \tag{2.13}$$

*holds for all trajectories* $\{u_k, y_k\}_{k=0}^{L-1}$ *of the system with initial condition* $x_0 = 0$, *where* $\{g_k^{\Psi}\}_{k=0,1,2,\ldots}$ *denotes the impulse response of* $\Psi$ *with* $g_k^{\Psi} \in \mathbb{R}^{n_r \times (m+p)}$.

*Proof.* The proof follows the arguments in the proof of Proposition 2.1. ∎

Dissipativity properties can be interpreted as one very important subclass of IQCs, where $\Psi$ and hence $P$ are constant matrices (e.g., $\Psi = I_{m+p}$, $M = \Pi \in \mathbb{R}^{(m+p)\times(m+p)}$). This subclass includes important input-output properties such as the $\mathcal{L}_2$-gain, (input and output strict) passivity, and conic relations (cf. Equation (2.5)).

Based on the problem setup, definitions, and results stated in this chapter, we will present three different approaches to determine input-output system properties from data in the following three chapters.

# Chapter 3

# Iterative schemes to determine input-output properties

The goal in this chapter is to present a systematic approach to iteratively determine certain dissipativity properties from input-output samples, where the input-output map remains undisclosed. In particular, we use multiple input-output trajectories to investigate the $\mathcal{L}_2$-gain, the shortage of passivity, and conic relations, respectively. The basic idea of iterative schemes is to adapt the input applied to the system based on the measured data in order to receive the system property of interest. By iteratively performing experiments, i.e., sampling input-output tuples, the input-output system property is then asymptotically revealed. This general idea is illustrated in Figure 3.1.

Previous work in this direction exploited the power method for iteratively determining the $\mathcal{L}_2$-gain of a system (e.g., [72, 82, 105, 106]). In this chapter, we broaden this approach to a more general framework on the basis of the following underlying ansatz.

**Sampling strategy for data-driven inference of input-output system properties:**

1. Formulate the system property as an optimization problem.
2. Apply the iterative sampling scheme:
   a) Compute the gradient of the formulated optimization problem from data.
   b) Update the input along the computed gradient.
   c) Perform experiments with the updated input.

This general approach, as schematically illustrated in Fig. 3.2, is hence based on a gradient scheme, where the exact cost function of the optimization problem depends on the underlying system and is hence a priori unknown. However, we will show that the gradients can be evaluated from only input-output data samples, which allows to employ gradient dynamical systems and saddle point flows to solve the reformulated optimization problems that yield the system property of interest.

Employing this idea, we present iterative sampling schemes, which reveal the $\mathcal{L}_2$-gain, passivity properties, or the cone with minimal radius containing the input-output behavior. We start in Section 3.1 with discrete-time SISO LTI systems and provide convergence results of the respective evolution equations for these three input-output system properties. Furthermore, we introduce advanced sampling schemes which can improve data efficiency. In Section 3.2, we generalize the framework presented in Section 3.1 by considering continuous-time systems in Section 3.2.1 and MIMO systems in Section 3.2.2. Additionally, we provide results on the robustness of the presented framework to measurement noise in Section 3.2.3. Finally, we apply the introduced approaches to different simulation examples in Section 3.3 and end with a short summary in Section 3.4.

This chapter is is based on and taken in parts literally from [AK8][1], [AK19][2], [AK21][3], [AK22][4].

## 3.1 Gradient-based samplings schemes

Since the premise is to determine system properties from input-output data, one natural approach is the input-output framework introduced and presented for example in [22] and [113]. Hence, we take the viewpoint that System (2.1) is an operator that maps inputs $u$ to outputs $y$. While this input-output map is often undisclosed in practical applications, we can perform simulations or experiments where we choose the input $u$ and measure the corresponding output $y$.

[1] A. Koch, J.M. Montenbruck, F. Allgöwer. "Sampling strategies for data-driven inference of input-output system properties." In: *IEEE Trans. Automat. Control* 66.3 (2021). pp. 1144–1159 © 2020 IEEE.

[2] A. Romer, J.M. Montenbruck, F. Allgöwer. "Data-driven inference of conic relations via saddle-point dynamics." In: *Proc. 9th IFAC Symp. Robust Control Design* (2018). pp. 586–591 © 2018 IFAC.

[3] A. Romer, J.M. Montenbruck, F. Allgöwer. "Sampling strategies for data-driven inference of passivity properties." In: *Proc. 56th IEEE Conf. on Decision and Control* (2017). pp. 6389–6394 © 2017 IEEE.

[4] A. Romer, J.M. Montenbruck, F. Allgöwer. "Some ideas on sampling strategies for data-driven inference of passivity properties for MIMO systems." In: *Proc. American Control Conference* (2018). pp. 6094–6100 © 2018 AACC.

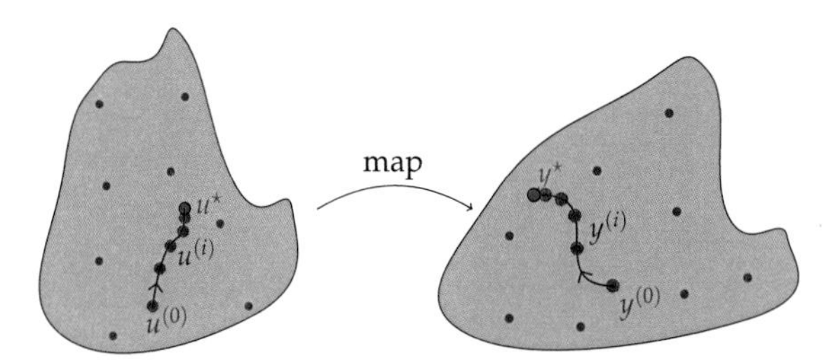


**Figure 3.1.** The research objective of iterative analysis schemes is to find input-output properties by drawing further data tuples $(u^{(i)}, y^{(i)})$ in order to converge to the input-output trajectory $(u^\star, y^\star)$ that corresponds to the respective system property, e.g., the input trajectory $u^\star$ for which System (2.1) exhibits the maximal gain from input to output, i.e., the $\mathcal{L}_2$-gain.

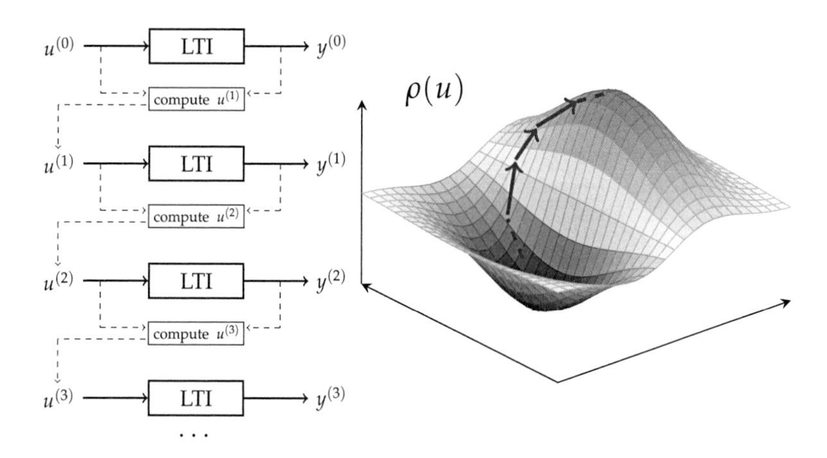


**Figure 3.2.** The general idea is to iteratively perform (numerical) experiments to converge to the parameter corresponding to a certain system property of System (2.1), e.g., the $\mathcal{L}_2$-gain. The gradient ascent algorithm is based on only input-output data while the system and hence the optimization function $u \mapsto \rho(u)$ remains undisclosed.

We start by considering the system class described in (2.1) restricted to $p = m = 1$, i.e., discrete-time SISO LTI systems. The input-output map of such systems can be described by

$$y_k = \sum_{l=0}^{\infty} g_l u_{k-l}, \tag{3.1}$$

where $\{g_k\}_{k=0,1,\ldots}$, $g_k \in \mathbb{R}$, denotes the impulse response sequence. For a given finite-time input sequence $\{u_k\}_{k=0}^{N-1}$, the input-to-output operator in (3.1) can be written in matrix notation

$$\begin{bmatrix} y_0 \\ \vdots \\ y_{N-1} \end{bmatrix} = \begin{bmatrix} g_0 & 0 & 0 & \ldots & 0 \\ g_1 & g_0 & 0 & \ldots & 0 \\ g_2 & g_1 & g_0 & \ldots & 0 \\ \vdots & \vdots & & \ddots & \vdots \\ g_{n-1} & g_{n-2} & \cdots & g_1 & g_0 \end{bmatrix} \begin{bmatrix} u_0 \\ \vdots \\ u_{N-1} \end{bmatrix}, \tag{3.2}$$

which is denoted in the following by $y = Gu$ with $u, y \in \mathbb{R}^N$ and $G \in \mathbb{R}^{N \times N}$. The matrix $G$ representing the convolution operator for finite-length inputs $u \in \mathbb{R}^N$ is a lower triangular Toeplitz matrix. Note that we assume $u_k = 0$ for $k < 0$ and only consider causal, asymptotically stable systems. However, the ideas of [94] can be applied for converting the sampling strategies to closed-loop approaches, where pre-stabilizing controllers enable the application to unstable systems.

### 3.1.1 $\mathcal{L}_2$-gain

The small-gain theorem, as for example presented in [113], plays an important role in systems analysis, stability studies, and controller design. With the knowledge of an upper bound on the $\mathcal{L}_2$-gain of open-loop elements, the stability of the closed loop can be validated. The $\mathcal{L}_2$-gain of System (2.1) is the smallest $\gamma$ for which the dissipativity condition in Definition 2.4 holds for the supply rate $s$ with $\Pi = \Pi_\gamma$ from (2.5). From the input-output viewpoint together with Theorem 2.2, this yields the condition that

$$\|y\| \leq \gamma \|u\| \tag{3.3}$$

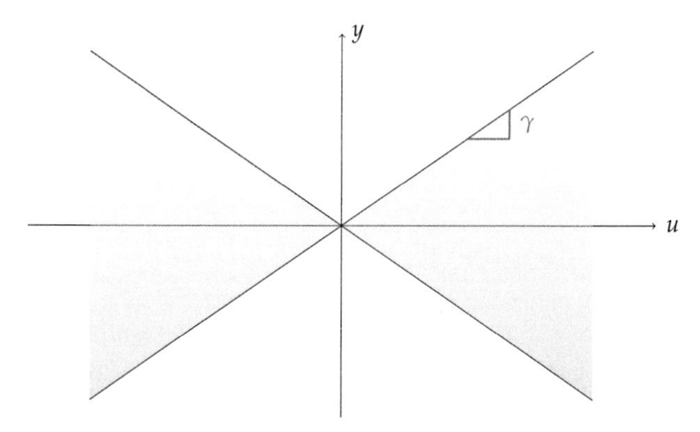


**Figure 3.3.** A graphical illustration of the $\mathcal{L}_2$-gain $\gamma$ in the input-output plane.

holds for all trajectories $\{u_k, y_k\}_{k=0}^h$ of System (2.1) with input $u \in l_2^m$ and initial condition $x_0 = 0$, and for all $h \geq 0$. A graphical interpretation of such a gain bound is depicted in Figure 3.3.

While the condition in (3.3) needs to hold for inputs and outputs of all lengths $h \geq 0$ to find the $\mathcal{L}_2$-gain, we will always only have access to finite-length data trajectories and can only perform experiments over a finite horizon $N < \infty$. Therefore, we will in the following consider the $\mathcal{L}_2$-gain (as well as passivity and conic relations) over a fixed horizon $L$ (cf. $L$-dissipativity in Definition 2.5), as also done in the related literature (e.g., [72, 82, 105, 106]). In the remainder of this chapter, the horizon over which we determine the respective system property is equal to the length of the available input-output trajectories $\{u_k, y_k\}_{k=0}^{N-1}$, i.e., $L = N$. To clarify this in the notation, we use $\gamma_N$ to denote the $\mathcal{L}_2$-gain over the horizon $N$, i.e., the minimum $\gamma$ such that (3.3) holds for all trajectories $\{u_k, y_k\}_{k=0}^h$ of System (2.1) with input $u \in l_2^m$ and initial condition $x_0 = 0$, and for all $0 \leq h \leq N$. As already shown in [82, 98], for $N \to \infty$, we naturally retrieve the true gain in the sense that $\lim_{N\to\infty} \gamma_N = \gamma$. Discussions and results on bounding the difference of the $\mathcal{L}_2$-gain over a finite horizon and an infinite horizon can be found in Section 4.3.

For an iterative model-free approach to determine $\gamma_N$, we formulate the condition in (3.3) in terms of an optimization problem, searching for the input for which the system exhibits the maximal gain from input to output. To this end, we employ the input-output map of the considered system class over a finite horizon as described in (3.2) together with the result of Proposition 2.1, which yields the optimization problem

$$\gamma_N^2 = \max_{u \in \mathbb{R}^N,\, \|u\|^2 \neq 0} \frac{\|y\|^2}{\|u\|^2} = \max_{u \in \mathbb{R}^N,\, \|u\|^2 \neq 0} \rho_1(u), \quad \text{with } \rho_1(u) = \frac{u^\top G^\top G u}{\|u\|^2}, \tag{3.4}$$

where the term $\rho_1$ is also referred to as the Rayleigh quotient. The Rayleigh quotient is a smooth function $\rho_1 : \mathbb{R}^N \setminus \{0\} \to \mathbb{R}$ that is scale-invariant since $\rho_1(u) = \rho_1(\alpha u)$ holds for all scalars $\alpha \neq 0$. Therefore, it is sufficient to consider the Rayleigh quotient on the unit sphere $\mathbb{S}^{N-1} = \{u \in \mathbb{R}^N | \|u\| = 1\}$. The critical points and critical values of $\rho_1$ are the eigenvectors and eigenvalues of $G^\top G$, respectively, as shown for example in [32]. Thus, the maximum value of the Rayleigh quotient in (3.4) is exactly the maximum eigenvalue $\lambda_1$ of the symmetric matrix $G^\top G$. This relation is also referred to as the variational characterization of eigenvalues or as the Courant-Fischer-Weyl principle.

The first proposition recasts a results of [106] and states that the gradient of the Rayleigh quotient can in fact be computed by only sampling two input-output trajectories, which can be generated, for example, from simulations or experiments.

**Proposition 3.1.** *The gradient vector field of $\rho_1 : \mathbb{S}^{N-1} \to \mathbb{R}$ is given by*

$$\nabla \rho_1(u) = 2G^\top G u - 2\rho_1(u)u \tag{3.5}$$

*and can be computed by evaluating $u \mapsto Gu$ twice.*

*Proof.* We endow the unit sphere $\mathbb{S}^{N-1}$ with the standard Riemannian metric, i.e., the Riemannian metric induced from the embedding $\mathbb{S}^{N-1} \subset \mathbb{R}^N$. Hence, the gradient at $u \in \mathbb{S}^{N-1}$ is uniquely determined by

$$\nabla \rho_1(u) = \frac{2G^\top G u \cdot u^\top u - 2u^\top G^\top G u \cdot u}{(u^\top u)^2} = 2G^\top G u - 2\rho_1(u)u.$$

In order to compute $\nabla\rho_1(u)$ from evaluating $u \mapsto Gu$, we define the involutory permutation matrix

$$T_P = \begin{bmatrix} 0 & \dots & 0 & 1 \\ 0 & \dots & 1 & 0 \\ \vdots & \iddots & & \vdots \\ 1 & \dots & 0 & 0 \end{bmatrix}$$

with $T_P = T_P^{-1} \in \mathbb{R}^{N\times N}$. Note that the matrices $G$ and $G^\top$ are involutory conjugate since $T_P G^\top = G T_P$ holds. Hence, we can compose $G^\top u$ by $G^\top u = T_P G T_P u$. This finally leads to $\nabla\rho_1(u) = 2T_P G T_P G u - 2(u^\top T_P G T_P G u)u$, which only consists of operations we can perform by evaluating $u \mapsto Gu$. ■

In experiments or simulations, the term $T_P G T_P G u$ can be obtained by performing one (numerical) experiment $y = Gu$, applying the reversed output $T_P G u$ to $G$ in a second experiment and reversing the output again.

In the following, we strive to solve (3.4) by applying gradient dynamical systems. This leads us to a continuous-time ordinary differential equation, which we then translate into a discrete-time optimization approach.

#### $\mathcal{L}_2$-gain - continuous-time solution

Since the gradient of the Rayleigh quotient can be evaluated by only performing experiments, we can find the maximum of the Rayleigh quotient without knowledge of the input-output map by employing a gradient dynamical system

$$\frac{\mathrm{d}}{\mathrm{d}\tau} u(\tau) = \nabla\rho_1(u(\tau)) \tag{3.6}$$

with $u(\tau) = \begin{bmatrix} u_0(\tau) & \cdots & u_{N-1}(\tau) \end{bmatrix}^\top$. As illustrated in Figure 3.4(a), $\rho_1$ increases monotonically along the solutions of (3.6). This leads us to the evolution equation

$$\frac{\mathrm{d}}{\mathrm{d}\tau} u(\tau) = 2G^\top G u(\tau) - 2\rho_1(u(\tau))u(\tau), \tag{3.7}$$

also known as the Rayleigh quotient gradient flow. It is readily verified that the gradient flow in (3.7) leaves the sphere $\mathbb{S}^{N-1}$ invariant [32].

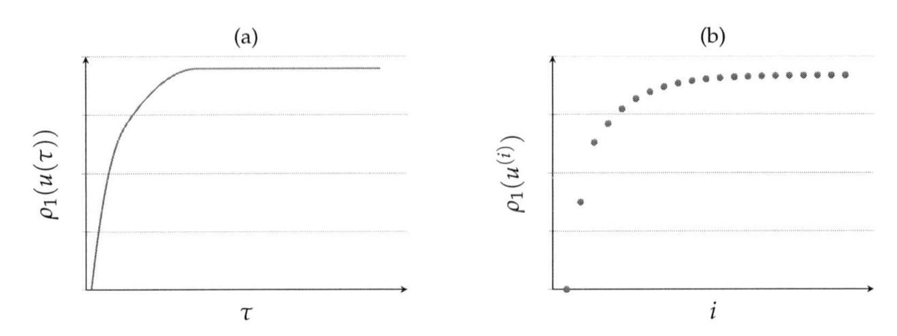


**Figure 3.4.** Illustration of the (a) continuous-time and (b) discrete-time gradient ascent optimization of the cost function $\rho_1 : \mathbb{S}^{N-1} \to \mathbb{R}$.

On the sphere $\mathbb{S}^{N-1}$, the Rayleigh quotient gradient flow is in fact equivalent to the so-called *Oja flow*

$$\frac{\mathrm{d}}{\mathrm{d}\tau} u(\tau) = (u(\tau)^\top (G^\top G - u(\tau)^\top G^\top G u(\tau) I_N) u(\tau), \tag{3.8}$$

defined in $\mathbb{R}^N$. The *Oja flow* is used, for example, in neural network learning theory as a means to determine the eigenvectors corresponding to the largest eigenvalues [70].

**Theorem 3.1.** *Assume $\lambda_1 > \lambda_2 \geq ... \geq \lambda_N$ for the eigenvalues $\lambda_i$ of $G^\top G$. For almost all initial conditions $u(0) \in \mathbb{R}^N$ with $\|u(0)\| = 1$, $\rho_1$ converges to $\gamma_N^2$ along the solutions of* (3.7).

*Proof.* This result follows directly from Theorem 3.4 in [32]. ■

### $\mathcal{L}_2$-gain - discrete-time solution

In an experimental setup, we can only iteratively determine the gradient from input-output data and hence, we extend the result to discrete-time optimization. Generally, an iterative approach for maximizing $\rho_1$ in (3.4) is to construct a sequence $u^{(i)}$, $i = 0, 1, 2, ...$, as illustrated in Figure 3.5, such that $\rho_1(u^{(i+1)}) > \rho_1(u^{(i)})$ holds for all $i$, as depicted in Figure 3.4(b).

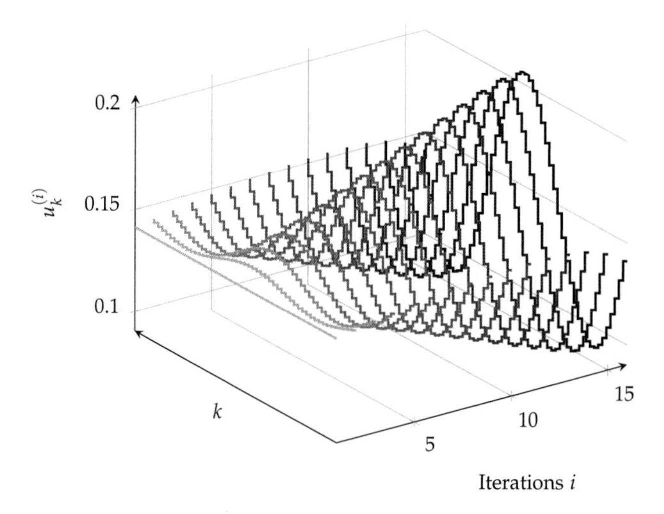


**Figure 3.5.** An inital input $u^{(0)} \in \mathbb{R}^N$ iteratively converges towards the input $u^\star$ which corresponds to the $\mathcal{L}_2$-gain given by $\gamma_N^2 = \rho_1(u^\star)$.

One standard tool in numerical linear algebra to construct such a sequence converging towards the dominant eigenvector and corresponding eigenvalue of a linear operator is the power method

$$u^{(i+1)} = \frac{G^\top G u^{(i)}}{\|G^\top G u^{(i)}\|}. \tag{3.9}$$

This method, which was amongst other methods proposed in [106], presents a possible approach for iteratively estimating the $\mathcal{L}_2$-gain without any knowledge of an explicit expression of $G$, since we can retrieve the expression $G^\top G u^{(i)}$ for any $u^{(i)} \in \mathbb{R}^N$ by sampling two input-output samples $G u^{(i)}$ and $G^\top G u^{(i)} = T_P G T_P G u^{(i)}$. Hence, by iterative input-output sampling, we can apply the power method for finding the $\mathcal{L}_2$-gain while the input-output operator remains undisclosed.

**Proposition 3.2** ([32, 106]). *Let the largest eigenvalue of $G^\top G$ be unique. Then, for almost all initial conditions $u^{(0)}$ with $\|u^{(0)}\| = 1$, the sequence $u^{(i)} \in \mathbb{R}^N$, $i = 0, 1, 2, \ldots$, constructed by (3.9) converges to the dominant eigenvector of $G^\top G$ and hence, $\rho_1$ converges to $\gamma_N^2$ along this sequence.*

With the scale-invariance of the Rayleigh quotient, the relevant information is contained in the direction of $u^{(i)}$. In other words, each element $u^{(i)} \in \mathbb{S}^{N-1}$ of the sequence represents the one-dimensional subspace $\{\beta u^{(i)} : \beta \in \mathbb{R}\}$. This links the presented approaches to optimizing over the real projective $(N-1)$-space, usually denoted by $\mathbb{RP}^{N-1}$. For further information, the interested reader is referred to [32].

**Remark 3.1.** The power iteration can actually be interpreted as a discrete-time version of the Rayleigh quotient gradient flow presented in (3.7) [49]. To this end, let us define $\pi_u : \mathbb{R}^n \to \mathbb{R}^n$ with $\pi_u(z) = z - (u^\top z)u$, which is a projection of $\mathbb{R}^N$ onto the tangent space $T_u\mathbb{S}^{N-1}$ at the point $u \in \mathbb{S}^{N-1}$. The continuous-time flow associated with the power method then reads

$$\begin{aligned}\frac{\mathrm{d}}{\mathrm{d}\tau}u(\tau) = \pi_{u(\tau)}\left(G^\top G u(\tau)\right) &= G^\top G u(\tau) - u(\tau)^\top G^\top G u(\tau) u(\tau)\\ &= (G^\top G - \rho_1(u(\tau)) I_N)u = \frac{1}{2}\nabla\rho_1(u(\tau)),\end{aligned}$$

retrieving the gradient dynamical system from before.

**Remark 3.2.** There is another well-known method that maximizes the Rayleigh quotient, namely the Rayleigh quotient iteration. The Raleigh quotient iteration uses the iteration scheme

$$u^{(i+1)} = \frac{((G^\top G) - \rho_1(u^{(i)}) I_N)^{-1} u^{(i)}}{\|((G^\top G) - \rho_1(u^{(i)}) I_N)^{-1} u^{(i)}\|} \tag{3.10}$$

to find the dominant eigenvector of $G^\top G$. The Rayleigh quotient iteration has stronger convergence properties than the power iteration [5], but it cannot be applied in the present setting since we cannot compute the right hand side without explicit knowledge of $G$.

To improve the convergence rate of (3.9), the application of the Lanczos method is introduced in [106]. In [82], Rojas et al. propose another approach to decrease the required input-output samples. Since $G$ is a lower triangular Toeplitz matrix, the matrix $T_{\mathrm{P}}G$ is symmetric and can be factorized into $Q_{\mathrm{o}}\Lambda Q_{\mathrm{o}}^\top$ where $Q_{\mathrm{o}}$ is an orthonormal matrix and $\Lambda$ is a diagonal matrix with the eigenvalues of $T_{\mathrm{P}}G$. With $G^\top G = T_{\mathrm{P}}GT_{\mathrm{P}}G = Q_{\mathrm{o}}\Lambda Q_{\mathrm{o}}^\top Q_{\mathrm{o}}\Lambda Q_{\mathrm{o}}^\top = Q_{\mathrm{o}}\Lambda^2 Q_{\mathrm{o}}^\top$, we hence find that the maximum absolute eigenvalue of $T_{\mathrm{P}}G$ is exactly the square root of the maximal eigenvalue of

$G^\top G$. Hence, finding the $\mathcal{L}_2$-gain by applying the power method to the matrix $T_P G$ requires only one sample per iteration.

### 3.1.2 Passivity

Besides the $\mathcal{L}_2$-gain, passivity is one of the key properties that can be exploited in order to analyze stability and design controllers, cf. [99]. The relevance of passivity for feedback control was recognized early, providing well-known feedback theorems for passive systems (cf. [22, 113]). System (2.1) is said to be passive if the dissipativity condition in Definition 2.4 holds for the supply rate $s$ with $\Pi = \begin{bmatrix} 0 & I \\ I & 0 \end{bmatrix}$. With Theorem 2.2, an alternative condition for passivity is that

$$\sum_{k=0}^{h} u_k^\top y_k \geq 0, \quad \forall h \geq 0, \tag{3.11}$$

holds for all input-output trajectories $\{u_k, y_k\}_{k=0}^{\infty}$ of System (2.1) with $u \in l_2^m$ and initial condition $x_0 = 0$. For controller design, however, we are specifically interested to which extent a system is or is not passive. The shortage of passivity is defined as the smallest $\beta \in \mathbb{R}$ such that

$$\sum_{k=0}^{h} u_k^\top y_k + \beta \|y_k\|^2 \geq 0, \quad \forall h \geq 0, \tag{3.12}$$

holds for all input-output trajectories $\{u_k, y_k\}_{k=0}^{\infty}$ of System (2.1) with $u \in l_2^m$ and initial condition $x_0 = 0$. The system is said to be output strictly passive if $\beta < 0$. A graphical illustration of output strict passivity is given in Fig. 3.6. For $\beta > 0$, the shortage of passivity corresponds to the required excess of passivity of a controller to render the closed loop stable. For a more detailed description of passivity and its relevance for the application of well-known feedback theorems, the reader is referred to [99], [113], and Chapter 6 of [22]. Another parameter to determine to which extent a system is or is not passive is the input feedforward passivity index, which is defined as the largest $\nu$ such that

$$\sum_{k=0}^{h} u_k^\top y_k - \nu \|u_k\|^2 \geq 0, \quad \forall h \geq 0,$$

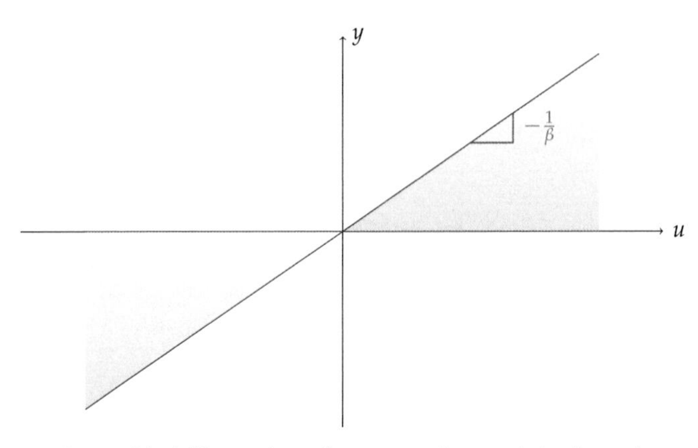


**Figure 3.6.** A graphical illustration of output strict passivity in a plane, where $\beta$ denotes the shortage of passivity.

holds for all input-output trajectories $\{u_k, y_k\}_{k=0}^{\infty}$ of System (2.1) with $u \in l_2^m$ and initial condition $x_0 = 0$. Hereby, the input feedforward passivity index $|\nu|$ corresponds to the minimum feedforward action required for rendering the system passive in the case of $\nu < 0$. If $\nu > 0$, then the system is said to be input strictly passive. While we focus on the shortage of passivity in the following, input strict passivity nicely fits into the general framework presented in this chapter and the extensions and discussions in Section 3.2 hence also hold for the input feedforward passivity index.

As in the case of the $\mathcal{L}_2$-gain, we only have access to trajectories of finite length in practice, and we will hence consider the shortage of passivity over a finite horizon $L$ (cf. $L$-dissipativity in Definition 2.5). Since the considered horizon is equal to the length $N$ of the available trajectories, we will in the following denote the shortage of passivity over the horizon $N$ by $\beta_N$ (and $\nu_N$ in the case of the input feedforward passivity index). Employing the input-output map of the considered system class over a finite horizon $N$ as given in (3.2) together with the result from Proposition 2.1, the Inequality (3.12) to retrieve the shortage of passivity reads

$$u^\top G u \geq -\beta_N u^\top G^\top G u, \tag{3.13}$$

which must hold for all inputs $u \in \mathbb{R}^N$. Let us now assume that $g_0 \neq 0$. Consequently, the Toeplitz matrix $G$ has full rank and $G^\top G$ is positive definite. Reformulating the definition of the system property in (3.13) in terms of an optimization problem, with $u^\top Gu = \frac{1}{2}u^\top(G+G^\top)u$ due to the symmetry in quadratic terms, leads to

$$\beta_N = -\min_{u\in\mathbb{R}^N,\ \|u\|\neq 0} \rho_2(u) = -\min_{u\in\mathbb{R}^N,\ \|u\|\neq 0} \frac{1}{2}\frac{u^\top(G+G^\top)u}{u^\top G^\top Gu}. \tag{3.14}$$

The term $\rho_2$ here is also referred to as the generalized Rayleigh quotient, which is a smooth function $\rho_2 : \mathbb{R}^N\backslash\{0\} \to \mathbb{R}$. This minimization problem represents a generalized eigenvalue problem, where the critical points and critical values of $\rho_2$ are the generalized eigenvectors $v_i$ and generalized eigenvalues $\lambda_i$ of the pair $(\frac{1}{2}(G+G^\top), G^\top G)$ defined by

$$\frac{1}{2}(G+G^\top)v_i = \lambda_i G^\top G v_i, \quad i = 1,\ldots,N.$$

Therefore, we are searching for the smallest generalized eigenvalue of the generalized eigenvalue problem denoted by $\lambda_N$ in order to retrieve the shortage of passivity $\beta_N = -\lambda_N$.

Due to the scale invariance of the generalized Rayleigh quotient $\rho_2$, we again consider the optimization problem in (3.14) on the sphere $\mathbb{S}^{N-1}$. The following proposition states that the gradient of the generalized Rayleigh quotient can be computed by only sampling three input-output tuples.

**Proposition 3.3.** *The gradient vector field of $\rho_2 : \mathbb{S}^{N-1} \to \mathbb{R}$ is given by*

$$\begin{aligned}\nabla\rho_2(u) &= \frac{1}{\|Gu\|^2}((G+G^\top)u - 2\rho_2(u)G^\top Gu) \\ &= \frac{Gu + T_P G T_P u}{\|Gu\|^2} - \frac{u^\top(Gu + T_P G T_P u)}{\|Gu\|^4}(T_P G)^2 u.\end{aligned} \tag{3.15}$$

*and can be computed by evaluating $u \mapsto Gu$ thrice.*

*Proof.* This proof follows the proof of Proposition 3.1. ∎

Computing the gradient vector field given in (3.15) hence requires the three data samples $(u, Gu)$, $(u, T_P G T_P u)$, and $(u, (T_P G)^2 u)$ from three consecutive (numerical) experiments. For reasons of measurement noise, it is recommendable to calculate $\|Gu\|^2$ by $u^\top\left((T_P G)^2 u\right)$ (cf. [106]).

**Passivity - continuous-time solution**

In order to find the smallest generalized eigenvalue $\lambda_N$ and hence the shortage of passivity, we employ the gradient dynamical system

$$\frac{\mathrm{d}}{\mathrm{d}\tau}u(\tau) = -\nabla\rho_2(u(\tau)) = \frac{1}{\|Gu(\tau)\|^2}(2\rho_2(u(\tau))G^\top Gu(\tau) - (G+G^\top)u(\tau)) \quad (3.16)$$

along whose solutions $\rho_2$ decreases monotonically. By

$$\frac{\mathrm{d}}{\mathrm{d}\tau}\|u(\tau)\|^2 = \frac{2}{\|Gu(\tau)\|^2}\left(2\rho_2(u(\tau))u(\tau)^\top G^\top Gu(\tau) - u(\tau)^\top(G+G^\top)u(\tau)\right) = 0,$$

we verify that (3.16) leaves the sphere $\mathbb{S}^{N-1}$ invariant.

Gradient flows of Morse-functions or more generally of Morse-Bott functions have, with the topological restrictions, strong convergence properties on manifolds. In order to prove convergence of $\rho_2$ to the shortage of passivity along the solutions of (3.16), we will therefore first show that $\rho_2$ on the unit sphere $\mathbb{S}^{N-1}$ is a Morse-Bott function, which we define in the following.

Let $\mathcal{M}$ be a smooth manifold and let $\Phi : \mathcal{M} \to \mathbb{R}$ be a smooth function. We say $\Phi$ is a Morse-Bott function provided the following three conditions from [32, p. 21] are satisfied:

a) $\Phi : \mathcal{M} \to \mathbb{R}$ has compact sublevel sets.

b) $C(\Phi) = \cup_{j=1}^{K}\mathcal{N}_j$ with $\mathcal{N}_j$ being disjoint, closed, and connected submanifolds of $\mathcal{M}$ and $\Phi$ being constant on $\mathcal{N}_j$, $j = 1, \ldots, K$.

c) $\ker(\mathrm{H}_\Phi(u)) = T_u\mathcal{N}_j$, for all $u \in \mathcal{N}_j$, $j = 1, \ldots, K$.

Here, $C(\Phi)$ denotes the set of critical points of $\Phi$, $\mathrm{H}_\Phi(u)$ denotes the Hessian of $\Phi$ at $u$, $\ker(\mathrm{H}_\Phi(u))$ denotes the kernel of the Hessian of $\Phi$ at $u$, and $T_u\mathcal{N}_j$ is the tangent space of $\mathcal{N}_j$ at $u$.

Due to the strong convergence properties of gradient flows of Morse-Bott functions, we show in the following lemma that $\rho_2 : \mathbb{S}^{N-1} \to \mathbb{R}$ is indeed a Morse-Bott function as defined above.

**Lemma 3.1.** *The generalized Rayleigh quotient $\rho_2$ on the unit sphere $\mathbb{S}^{N-1}$ is a Morse-Bott function.*

*Proof.* Condition a) of the definition of a Morse-Bott function requires that for all $c \in \mathbb{R}$ the sublevel set $\{u \in \mathbb{S}^{N-1} | \rho_2(u) \leq c\}$ is a compact subset of $\mathbb{S}^{N-1}$. Since $\mathbb{S}^{N-1}$ is compact and $\rho_2$ is continuous, this is satisfied.

The critical points $C(\rho_2)$ are all $u \in \mathbb{S}^{N-1}$ such that

$$\left(\frac{1}{2}(G+G^\top) - \rho_2(u)G^\top G\right) u = 0,$$

which are exactly the generalized eigenvectors of the matrix pair $(\frac{1}{2}(G+G^\top), G^\top G)$. All eigenvalues with geometric multiplicity one are hence isolated critical points.

If the generalized eigenvalue $\lambda_i$ has geometric multiplicity of $m$, then the solution of the equation

$$\left(\frac{1}{2}(G+G^\top) - \lambda_i G^\top G\right) u = 0$$

is an $m$-dimensional linear subspace of $\mathbb{R}^N$ and a closed connected submanifold $\mathcal{N}_i$ of dimension $m-1$ on the unit sphere $\mathbb{S}^{N-1}$. On this submanifold, $\rho_2(u) = \lambda_i$ for all $u \in \mathcal{N}_i$. Therefore, condition b) is also satisfied.

Finally, we need to show that also condition c) holds. According to the above discussion, $T_u\mathcal{N}_j$ is contained in $\ker\left(\mathtt{H}_{\rho_2}(u)\right)$, and we only have to show that $\ker\left(\mathtt{H}_{\rho_2}(u)\right) \subseteq T_u\mathcal{N}_j \ \forall u \in \mathcal{N}_j$. Hence, we start by calculating the Hessian of $\rho_2$ at the critical points.

Let $v_i \in C(\rho)$ be the generalized eigenvector corresponding to the generalized eigenvalue $\lambda_i$ of multiplicity $m$. Then the symmetric Hessian matrix of $\rho_2$ at $v_i$ is given by

$$\mathtt{H}_{\rho_2}(v_i) = \frac{2}{\|Gv_i\|^2}\left(\frac{1}{2}(G+G^\top) - \lambda_i G^\top G\right).$$

In this case, the vector $v_i$ is an element of an $(m-1)$-dimensional submanifold $\mathcal{N}_i$, and the nullspace of the Hessian $\mathtt{H}_{\rho_2}(v_i)$ is exactly the eigenspace corresponding to $\lambda_i$. Let $\psi \in T_{v_i}\mathbb{S}^{N-1}$ have a normal component to the eigenspace $\mathcal{N}_i$, but $\psi$ still lies in the kernel of the Hessian $\mathtt{H}_{\rho_2}(v_i)$. Then

$$\lambda_i G^\top G\psi - \frac{1}{2}(G+G^\top)\psi = 0$$

must hold, $(\psi, \lambda_i)$ is thus a solution to the generalized eigenvalue problem, and $\psi \in \mathcal{N}_i$, which leads to a contradiction. Hence, the Hessian $\mathrm{H}_{\rho_2}(v_i)$ has full rank in any direction normal to $\mathcal{N}_i$ at any $v_i \in \mathcal{N}_i$. We say that every critical point of $\mathbb{S}^{N-1}$ belongs to a nondegenerate critical submanifold.

Altogether, we have shown that the generalized Rayleigh quotient $\rho_2$ is indeed a Morse-Bott function on the unit sphere $\mathbb{S}^{N-1}$, which concludes the proof. ■

With this result, we find strong convergence guarantees for the generalized Rayleigh quotient flow, summarized in the following theorem.

**Theorem 3.2.** *Assume $\lambda_N < \lambda_{N-1} \leq \cdots \leq \lambda_1$ for the generalized eigenvalues $\lambda_i$ of the matrix pair $\left(\frac{1}{2}(G+G^\top), G^\top G\right)$. For almost all initial conditions $u^{(0)}$ with $\|u^{(0)}\| = 1$, $\rho_2$ converges to $-\beta_N$ along the solutions of* (3.16).

*Proof.* We start by showing that $\rho_2$ has two minima at the eigenvector $\pm v_N$ corresponding to the smallest eigenvalue $\lambda_N$. All other critical points are saddle points or maxima of $\rho_2$.

The linearization of (3.16) on the unit sphere at any generalized eigenvector $v_i$ corresponding to the generalized eigenvalue $\lambda_i$, $i = 1, \dots N$, reads

$$\frac{\partial}{\partial \tau} u(\tau) = \frac{2}{\|Gv_i\|^2} \left( \lambda_i G^\top G - \frac{1}{2}(G + G^\top) \right) u(\tau), \quad u(\tau)^\top v_i = 0.$$

To study the exponential stability of the critical points, we are now interested in the eigenvalues of the symmetric matrix $(\lambda_i G^\top G - \frac{1}{2}(G+G^\top))$ for all $i = 1, \dots, N$ corresponding to the critical points $v_i$. Since $(\lambda_i G^\top G - \frac{1}{2}(G+G^\top))$ is a symmetric matrix, the eigenvalues of $(\lambda_i G^\top G - \frac{1}{2}(G+G^\top))$ in the tangent space $T_{v_i}\mathbb{S}^{N-1}$ are all negative if and only if

$$u^\top \left( \lambda_i G^\top G - \frac{1}{2}(G + G^\top) \right) u < 0 \quad \forall u \in T_{v_i}\mathbb{S}^{N-1}. \tag{3.17}$$

From the definition of critical points, we know $\lambda_i G^\top G v_i - \frac{1}{2}(G+G^\top)v_i = 0$ for all $i = 1, \dots, N$. Adding $\lambda_N G^\top G v_i$ to both sides reads

$$\lambda_N G^\top G v_i - \frac{1}{2}(G + G^\top) v_i = (\lambda_N - \lambda_i) G^\top G v_i. \tag{3.18}$$

Since $\frac{1}{2}(G+G^\top)$ and $G^\top G$ are symmetric matrices, there exists a basis for $\mathbb{R}^N$ of generalized eigenvectors, which are $G^\top G$-orthogonal (i.e., $v_i^\top G^\top G v_j = 0$, for $i \neq j$).

Thus, every vector $u \in T_{v_i}\mathbb{S}^{N-1}$ can be decomposed into a linear combination of these generalized eigenvectors $v_i$, $i = 1, \ldots, N$. Multiplying $v_i^\top$ on both sides of (3.18), we retrieve

$$v_i^\top \left(\lambda_N G^\top G - \frac{1}{2}(G + G^\top)\right) v_i = (\lambda_N - \lambda_i)\|Gv_i\|^2$$

which is strictly less than zero for all $i \neq N$. With $u = \sum_{i=1}^{N} \alpha_i v_i$, where $\alpha_i \in \mathbb{R}$ for $i = 1, \ldots, N$, we find

$$u^\top \left(\lambda_N G^\top G - \frac{1}{2}(G + G^\top)\right) u = \sum_{i=1}^{N} \alpha_i^2 (\lambda_N - \lambda_i)\|Gv_i\|^2$$

which is negative if at least one $\alpha_i \neq 0$ for any $i{=}1, \ldots, N{-}1$. With the condition in (3.17), the eigenvalues of $(\lambda_N G^\top G - \frac{1}{2}(G + G^\top))$ in the tangent space $T_{v_N}\mathbb{S}^{N-1}$ are hence all negative, and the critical points $\pm v_N$ are exponentially stable. Analogously, the definition of generalized eigenvalues yields

$$v_N^\top \left(\lambda_i G^\top G - \frac{1}{2}(G + G^\top)\right) v_N = (\lambda_i - \lambda_N)\|Gv_N\|^2.$$

We now choose $\alpha \in \mathbb{R}$ such that $(v_N + \alpha v_i) \in T_{v_i}\mathbb{S}^{N-1}$. This yields

$$(v_N + \alpha v_i)^\top \left(\lambda_i G^\top G - \frac{1}{2}(G + G^\top)\right) (v_N + \alpha v_i) = (\lambda_i - \lambda_N)\|Gv_N\|^2,$$

which is strictly greater than zero for all $i \neq N$. With the condition in (3.17), there therefore exists at least one positive eigenvalue of $(\lambda_i G^\top G - \frac{1}{2}(G + G^\top))$ on the tangent space $T_{v_i}\mathbb{S}^{N-1}$ for all $i \neq N$. Any critical point $v_i$ with $i \neq N$ is hence a saddle point or a maximum of $\rho_2$.

Due to the reasoning above, only the isolated critical points $\pm v_N$ can be attractors for (3.16). With Lemma 3.1, $\rho_2$ is a Morse-Bott function and Proposition 3.9 from [32] applies. Therefore, every solution of the gradient flow converges to an equilibrium point as $\tau \to \infty$. Since the union of the unstable submanifolds corresponding to the generalized eigenspaces of the matrix pair $(\frac{1}{2}(G + G^\top), G^\top G)$ for the generalized eigenvalues $\lambda_1, \ldots \lambda_{N-1}$ forms a closed subset of co-dimension at least one, the solutions of (3.16) will converge to either $v_N$ or $-v_N$ for almost all initial conditions $u^{(0)}$ (cf. [32]). This completes the proof. ■

**Remark 3.3.** One other approach to investigate the convergence of the gradient dynamical system in (3.16) is to perform a linear coordinate transform $y = Gu$ to rewrite the generalized Rayleigh quotient $\rho_2$ into a standard Rayleigh quotient and then apply the results from [32] similar to the input strict passivity case below. Since $G$ is full rank, there always exists an inverse transformation $G^{-1}$. Define $T_R = G^{-\top}(G + G^\top)G^{-1}$. Using the transformation $y = Gu$, we retrieve the standard Rayleigh quotient

$$\rho_2(u) = \frac{1}{2}\frac{u^\top G^\top G^{-\top}(G + G^\top)G^{-1}Gu}{u^\top G^\top Gu} = \frac{1}{2}\frac{y^\top T_R y}{y^\top y}$$

with the symmetric matrix $T_R$, where the eigenvalues of $T_R$ correspond to the generalized eigenvalues $\lambda_i$ of the pair $(\frac{1}{2}(G + G^\top), G^\top G)$. With $G$ unknown, however, this transformation cannot directly be used for an iterative scheme to determine the shortage of passivity.

Similar strong convergence guarantees as in Theorem 3.2 can be given in the case of the input feedforward passivity index, where the optimization problem yields again a Rayleigh quotient similar to the $\mathcal{L}_2$-gain case. To find the input feedforward passivity index, however, we are searching for the smallest eigenvalue of an unknown matrix.

**Theorem 3.3.** *Assume $\lambda_N < \lambda_{N-1} \leq \cdots \leq \lambda_1$ for the eigenvalues $\lambda_i$ of the matrix $\frac{1}{2}(G + G^\top)$. For almost all initial conditions $u^{(0)}$ with $\|u^{(0)}\| = 1$,*

$$\frac{1}{2}\frac{u(\tau)^\top (G + G^\top)u(\tau)}{u^\top u}$$

*converges to $\nu_N$ along the solutions of*

$$\frac{\mathrm{d}}{\mathrm{d}\tau}u(\tau) = \frac{u(\tau)^\top (G + G^\top)u(\tau)}{u^\top u}u(\tau) - (G{+}G^\top)u(\tau). \tag{3.19}$$

*Proof.* This result follows directly from Theorem 3.4 in [32]. ∎

The gradient, i.e., the right-hand side of (3.19), can be determined by evaluating $u \mapsto Gu$ twice, as in the $\mathcal{L}_2$-gain case. For a more detailed discussion about this result on input strict passivity, the reader is referred to [AK21].

**Passivity - discrete-time solution**

In any application, we can only iteratively determine the gradient. We thus extend the previous results to discrete-time optimization where we improve the convergence through exact line search. Generally speaking, discrete-time minimization problems on manifolds can be approached by the general update formula

$$u^{(i+1)} = R_{u^{(i)}}(\alpha^{(i)} p^{(i)}) \tag{3.20}$$

where the search direction $p^{(i)}$ lies in the tangent space $T_{u^{(i)}}\mathbb{S}^{N-1}$ and $\alpha^{(i)}$ denotes the step size. The mapping $R_{u^{(i)}}$ is also called a retraction mapping from the tangent space $T_{u^{(i)}}\mathbb{S}^{N-1}$ to the manifold $\mathbb{S}^{N-1}$ [1]. Choosing

$$R_{u^{(i)}}(\alpha^{(i)} p^{(i)}) = \frac{u^{(i)} + \alpha^{(i)} p^{(i)}}{\|u^{(i)} + \alpha^{(i)} p^{(i)}\|} \tag{3.21}$$

yields a valid retraction onto the sphere $\mathbb{S}^{N-1}$ [1], which is defined for all vectors that lie in the tangent space $T_{u^{(i)}}\mathbb{S}^{N-1}$.

In the remainder of this section, we choose the search direction to be the negative gradient $p^{(i)} = -\nabla\rho_2(u^{(i)}) \in T_{u^{(i)}}\mathbb{S}^{N-1}$, which can be computed from data tuples according to Proposition 3.3. There exist various approaches on how to choose the step size $\alpha^{(i)}$. Literature on this topic has a long history and goes back to [34, 35], where the convergence for (generalized) Rayleigh quotient iterations with fixed step size or optimized step sizes are investigated. In fact, even though the input-output map of the discrete-time LTI system remains undisclosed, we can still perform a line search algorithm in the present setting. Minimizing

$$2\rho_2(u^{(i)} + \alpha^{(i)} p^{(i)}) = \frac{\begin{bmatrix} 1 \\ \alpha^{(i)} \end{bmatrix}^\top \begin{bmatrix} u^{(i)\top}(G+G^\top)u^{(i)} & u^{(i)\top}(G+G^\top)p^{(i)} \\ u^{(i)\top}(G+G^\top)p^{(i)} & p^{(i)\top}(G+G^\top)p^{(i)} \end{bmatrix} \begin{bmatrix} 1 \\ \alpha^{(i)} \end{bmatrix}}{\begin{bmatrix} 1 \\ \alpha^{(i)} \end{bmatrix}^\top \begin{bmatrix} u^{(i)\top}G^\top G u^{(i)} & u^{(i)\top}G^\top G p^{(i)} \\ u^{(i)\top}G^\top G p^{(i)} & p^{(i)\top}G^\top G p^{(i)} \end{bmatrix} \begin{bmatrix} 1 \\ \alpha^{(i)} \end{bmatrix}} \tag{3.22}$$

with respect to the step size $\alpha^{(i)}$ yields yet another generalized eigenvalue problem. Scaling the eigenvector that corresponds to the smaller eigenvalue such that its first entry equals one, its second entry denotes the optimized step size $\alpha_\star^{(i)}$. The optimized step size can again be computed by evaluating $u \mapsto Gu$ three additional times, without knowledge of $G$.

In the following, we show that the line search for the optimized step size $\alpha_\star^{(i)}$ can even be performed without any additional input-output samples. To this end, we generalize the main result from [95], which improves the sampling scheme for the input feedforward passivity index introduced in [AK21] by showing that no additional input-output samples are required for finding the optimized step size in the case of input strict passivity. Similarly, we show in the following that this also holds for the shortage of passivity via induction.

**Proposition 3.4.** *Given* $(G+G^\top)u^{(i)}$, $G^\top G u^{(i)}$, $(G+G^\top)p^{(i)}$, $G^\top G p^{(i)}$ *and* $\alpha_\star^{(i)}$. *Then the gradient* $p^{(i+1)}$ *and the optimal step size* $\alpha_\star^{(i+1)}$ *can be computed by evaluating* $u \mapsto Gu$ *thrice.*

*Proof.* Since $(G+G^\top)u^{(i+1)} = (G+G^\top)u^{(i)} + \alpha_\star^{(i)}(G+G^\top)p^{(i)}$ and $G^\top G u^{(i+1)} = G^\top G u^{(i)} + \alpha_\star^{(i)} G^\top G p^{(i)}$ holds, $p^{(i+1)}$ can be computed without additional input-output tuples. With the additional data tuples $(p^{(i+1)}, Gp^{(i+1)})$, $(T_\mathrm{P} p^{(i+1)}, G T_\mathrm{P} p^{(i+1)})$, and $(T_\mathrm{P} G p^{(i+1)}, G T_\mathrm{P} G p^{(i+1)})$, we can calculate the optimal step size $\alpha_\star^{(i+1)}$ via (3.22) and also fulfill the requirement to apply this proposition at step $i+1$. ∎

Further ideas on how to improve the data efficiency of the sampling scheme to find the input feedforward passivity index can be found in [43, AK11].

### 3.1.3 Conic relations

In [113], Zames introduces a feedback theorem on conic relations, which can be seen as a generalization of the small-gain theorem. In practice, the small-gain theorem provides often quite restrictive conditions for closed-loop stability. With a linear shift in the feedback equation, as illustrated in Figure 3.7, a reduced gain product can usually be obtained leading to a less conservative stability condition. This results in Zames' Theorem which says that the closed loop is bounded if the open loop can be factored into two, suitably proportioned, conic relations.

System (2.1) is said to be confined to a cone defined by the real constants $c$ and $r \geq 0$ if the inequality

$$\|y - cu\| \leq r\|u\| \tag{3.23}$$

holds for all trajectories $\{u_k, y_k\}_{k=0}^h$ of System (2.1) with input $u \in l_2^m$ and initial condition $x_0 = 0$, and for all $h \geq 0$. The constant $c$ is also called the center parameter

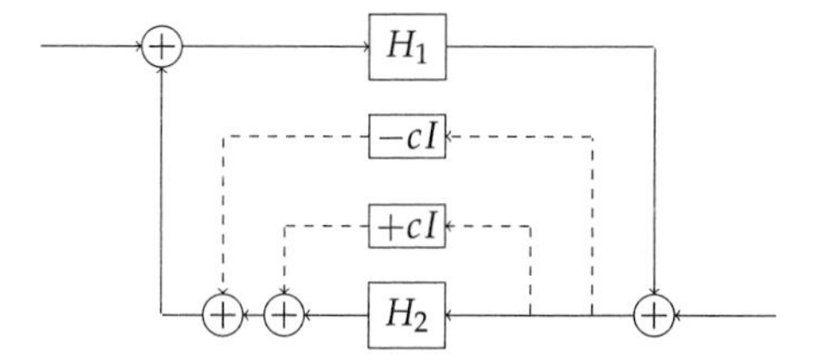


**Figure 3.7.** A transformation of a feedback loop with two elements [113].

and the constant $r$ is also called the radius of the input-output map. In Fig. 3.8, a graphical interpretation of such a conic sector in the plane is depicted.

Reformulating (3.23) in terms of an optimization problem yields

$$r^2 = \sup_{u \in l_2^m,\ \|u\|^2 \neq 0,\ h \geq 0} \frac{\sum_{k=0}^{h} \|y_k\|^2 - 2cu_k^\top y_k + c^2\|u_k\|^2}{\sum_{k=0}^{h} \|u_k\|^2}.$$

With this maximization problem, we can find a valid radius $r$ corresponding to any center $c$ describing a cone that System (2.1) is confined to. However, the goal is to find the transformation $\pm cI$ that minimizes the gain of the open-loop element, i.e., we strive to determine the center parameter $c$ which yields the minimum radius $r_{\text{min}}$ of the shifted unknown system. Finding $r_{\text{min}}$ can increase the set of controllers for which the closed loop is bounded. Equivalently, minimizing the radius can offer higher robustness measures for a given stabilizing controller related to the gap metric. As presented in [30] and [84], the gap between the cones, to which the open-loop elements are confined, can be interpreted as a robustness measure.

Searching for the minimum radius $r_{\text{min}}$ leads to

$$r_{\text{min}}^2 = \inf_{c \in \mathbb{R}} \sup_{u \in l_2^m,\ \|u\|^2 \neq 0,\ h \geq 0} \frac{\sum_{k=0}^{h} \|y_k\|^2 - 2cu_k^\top y_k + c^2\|u_k\|^2}{\sum_{k=0}^{h} \|u_k\|^2}, \tag{3.24}$$

which is a min-max optimization problem in the variables $c$ and $u$. Considering again only input-output trajectories $\{u_k, y_k\}_{k=0}^{N-1}$ of finite length $N$, we will in the following denote $r_{N,\text{min}}$ as the minimum radius of a conic relation over all centers $c$ that the unknown system satisfies over the horizon $N$ (cf. $L$-dissipativity in Definition 2.5). Employing the input-output map of the considered system class over a finite horizon $N$ as given in (3.2) together with the result of Proposition 2.1, we can rewrite the

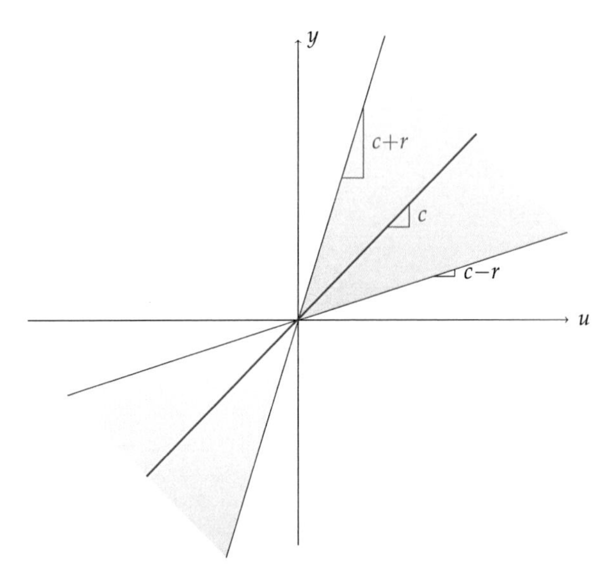


**Figure 3.8.** A graphical illustration of a conic sector in a plane [113], which is described by the center parameter $c$ and the radius $r$.

optimization problem from (3.24) into

$$r^2_{N,\min} = \min_{c\in\mathbb{R}} \max_{u\in\mathbb{R}^N,\|u\|\neq 0} \rho_3(c,u) = \min_{c\in\mathbb{R}} \max_{u\in\mathbb{R}^N,\|u\|\neq 0} \frac{u^\top(G^\top G - c(G+G^\top) + c^2 I_N)u}{\|u\|^2}. \tag{3.25}$$

The term $\rho_3 : \mathbb{R} \times \mathbb{R}^N \backslash \{0\} \to \mathbb{R}$ in (3.25) with $F(c) = G^\top G - c(G+G^\top) + c^2 I_N$ can again be referred to as a Rayleigh quotient. The Rayleigh quotient $\rho_3$ is a smooth function, which is scale-invariant in $u$. Therefore, we consider the Rayleigh quotient $\rho_3$ on the manifold $\mathbb{R} \times \mathbb{S}^{N-1}$.

Due to the Courant-Fischer-Weyl principle, all critical points and critical values of $\rho_3$ for any given $c$ are the eigenvectors and eigenvalues of $F(c)$, respectively. More specifically, the maximum of the Rayleigh quotient for any given $c$ corresponds to the largest eigenvalue $\lambda_1$ of $F(c)$. Hence, (3.25) can also be expressed as

$$r^2_{N,\min} = \min_{c\in\mathbb{R}} \lambda_1(F(c)). \tag{3.26}$$

This reveals that we are searching for the minimization of the maximal eigenvalue of a symmetric matrix function. The first result states that (3.26) is a strongly convex function with exactly one minimum.

**Lemma 3.2.** *The function $f_1 : \mathbb{R} \to \mathbb{R}$ with $c \mapsto \lambda_1(F(c))$ is strongly convex and has only one minimum, which is a global minimum.*

*Proof.* We start by separating $f_1(c) = \lambda_1(G^\top G - c(G + G^\top)) + c^2$. While $c^2$ is a strongly convex function, we are further interested in $\lambda_1(G^\top G - c(G+G^\top))$. From [26, Theorem A.1], the largest eigenvalue is convex and continuous in the space of symmetric matrices. Hence, the largest eigenvalue of an affine function of symmetric matrices $\lambda_1(G^\top G - c(G+G^\top))$ is convex. Since $c^2$ is a smooth and strongly convex function and $\lambda_1(G^\top G - c(G + G^\top))$ is continuous and convex, their sum is strongly convex and continuous. Hence, the function $f_1 : \mathbb{R} \to \mathbb{R}$ with $c \mapsto \lambda_1(F(c))$ has a global minimizer and no other local minima. ■

To find the minimal radius $r_{N,\min}$ from the min-max optimization problem in (3.25) without knowledge of $G$, we again apply a gradient-based optimization scheme. Therefore, the first proposition states that we can indeed retrieve the gradient of $\rho_3$ with respect to $c$ and $u$ from drawing input-output samples from simulations or experiments.

**Proposition 3.5.** *The gradients of $\rho_3 : \mathbb{R} \times \mathbb{S}^{N-1} \to \mathbb{R}$ in the first and second variable are given by*

$$\nabla_c \rho_3(c, u) = 2c - \frac{u^\top (G + G^\top) u}{\|u\|^2} = 2c - u^\top (Gu + T_P G T_P u), \tag{3.27}$$

$$\begin{aligned} \nabla_u \rho_3(c, u) &= \frac{2}{\|u\|^2}(F(c) - \rho_3(c, u) I_N) u \\ &= 2(T_P G T_P G u - c(Gu + T_P G T_P u)) - 2u^\top (T_P G T_P G u - c(Gu + T_P G T_P u))u, \end{aligned} \tag{3.28}$$

*and can be computed by evaluating $u \mapsto Gu$ thrice.*

*Proof.* This proof follows the proof of Proposition 3.1. ■

### Conic relations - continuous-time solution

To find the conic relation with minimal radius $r_{N,\min}$ for the unknown input-output system, we employ a gradient descent in the first variable $c$ and a gradient ascent in the second vector variable $u$ resulting in the saddle point dynamics given by

$$\begin{aligned} \frac{\mathrm{d}}{\mathrm{d}\tau} c(\tau) &= -\nabla_c \rho_3(c(\tau), u(\tau)), \\ \frac{\mathrm{d}}{\mathrm{d}\tau} u(\tau) &= \quad \nabla_u \rho_3(c(\tau), u(\tau)). \end{aligned} \tag{3.29}$$

The saddle point dynamics in (3.29) leave the manifold $\mathbb{R} \times \mathbb{S}^{N-1}$ invariant since

$$\frac{\mathrm{d}}{\mathrm{d}\tau} \|u(\tau)\|^2 = \frac{4}{\|u\|^2} u(\tau)^\top (F(c(\tau)) - \rho_3(c(\tau), u(\tau)) I_N) u(\tau) = 0.$$

The equilibrium points of (3.29) are described by $u$ being an eigenvector $v_i$ of $F(c)$ and the corresponding $c = \frac{1}{2} u (G + G^\top) u$. With the analysis before, we are searching for $u^\star$ being the eigenvector corresponding to the maximum eigenvalue denoted by $u^\star = v_1(F(c^\star))$ with $c^\star = \frac{1}{2} u^{\star\top} (G + G^\top) u^\star$, which then leads to the minimal radius $r_{N,\min}^2 = \rho_3(c^\star, u^\star)$.

In the following theorem, we show that the tuple with the center $c^\star$ and input sample $u^\star$ corresponding to the minimal radius $r_{N,\min}$ is in fact a locally attracting equilibrium point of (3.29) when $\lambda_1(F(c^\star))$ is a simple eigenvalue. When we minimize the maximal eigenvalue of a matrix function, however, we also need to consider the possibility that the solution to this optimization problem is an eigenvalue of geometric multiplicity two. In this second case, let $v_1(F(c^\star))$ and $v_2(F(c^\star))$ be the eigenvectors to the eigenvalue $\lambda_1(F(c^\star))$. We choose $v_1(F(c^\star))$ such that $2c - v_1(F(c))^\top (G + G^\top) v_1(F(c)) = 0$ and $v_2(F(c^\star))$ consecutively such that $v_2(F(c^\star))^\top v_1(F(c^\star)) = 0$, which is always possible since $F(c)$ is a symmetric matrix. We formulate an additional assumption in the case that $\lambda_1(F(c^\star))$ is an eigenvalue of multiplicity two.

**Assumption 3.1.** *With $v_1(F(c^\star))$ and $v_2(F(c^\star))$ as defined above, the following condition holds:*

$$v_1(F(c^\star))^\top (G + G^\top) v_2(F(c^\star)) \neq 0.$$

We will briefly discuss this technical assumption after the following theorem.

**Theorem 3.4.** *Assume that $\lambda_1(F(c^\star))$ is an eigenvalue*

- *with multiplicity one, or*
- *with multiplicity two and Assumption 3.1 holds.*

*Then the equilibrium point $(c^\star, u^\star)$ corresponding to the squared minimum radius $r_{N,\min}^2 = \rho_3(c^\star, u^\star)$ is locally exponentially stable under the saddle point dynamics in (3.29).*

*Proof.* Linearizing (3.29) around the critical point $(c^\star, u^\star) \in \mathbb{R} \times \mathbb{S}^{N-1}$ yields the linearized system dynamics

$$\frac{\mathrm{d}}{\mathrm{d}\tau}\begin{bmatrix}\delta c(\tau)\\ \delta u(\tau)\end{bmatrix} = J(c^\star, u^\star)\begin{bmatrix}\delta c(\tau)\\ \delta u(\tau)\end{bmatrix}$$

with $\delta c = c - c^\star$, $\delta u = u - u^\star$, and the Jacobian reads

$$J(c^\star, u^\star) = \begin{bmatrix} -\nabla_{cc}\rho_3(c^\star, u^\star) & -\nabla_{cu}\rho_3(c^\star, u^\star) \\ \nabla_{uc}\rho_3(c^\star, u^\star) & \nabla_{uu}\rho_3(c^\star, u^\star)\end{bmatrix}$$

where

$$\begin{aligned}
-\nabla_{cc}\rho_3(c^\star, u^\star) &= -2,\\
-\nabla_{cu}\rho_3(c^\star, u^\star) &= 2\left((G+G^\top)u^\star - (u^{\star\top}(G+G^\top)u^\star)u^\star\right)^\top,\\
\nabla_{uc}\rho_3(c^\star, u^\star) &= \nabla_{cu}\rho_3(c^\star, u^\star)^\top,\\
\nabla_{uu}\rho_3(c^\star, u^\star) &= 2\left(F(c^\star) - \rho_3(c^\star, u^\star)I_N\right).
\end{aligned}$$

Since any symmetric matrix possesses $N$ mutually orthogonal eigenvectors, the set of eigenvectors of the symmetric matrix $\frac{1}{2}\left(J(c^\star, u^\star) + J(c^\star, u^\star)^\top\right)$ given by $b_1 = (1, 0_N)$, $b_2 = (0, v_N)$, $b_3 = (0, v_{N-1})$, …, $b_N = (0, v_2)$, $b_{N+1} = (0, v_1)$ form an orthonormal basis of $\mathbb{R}^{N+1}$, where $v_i$, $i = 1, \dots, N$ denote the eigenvectors of $F(c^\star)$. If $\lambda_1(F(c^\star))$ has multiplicity two, we choose the two eigenvectors spanning the eigenspace corresponding to $\lambda_1(F(c^\star))$ such that $v_1 = u^\star$ and $v_1^\top v_2 = 0$.

Recall that (3.29) leaves the manifold $\mathbb{R} \times \mathbb{S}^{N-1}$ invariant, and we are thus only interested in the Jacobian on the tangent space $T_{(c^\star,u^\star)}\left(\mathbb{R} \times \mathbb{S}^{N-1}\right)$, which is spanned by the basis vectors $b_1$, $b_2$, …, $b_N$. By projecting the Jacobian matrix $J(c^\star, u^\star)$ onto the tangent space $T_{(c^\star,u^\star)}\left(\mathbb{R} \times \mathbb{S}^{N-1}\right)$, we find

$$J'(c^\star, u^\star) = \begin{bmatrix} b_1^\top \\ \vdots \\ b_N^\top \end{bmatrix} J(c^\star, u^\star)\begin{bmatrix} b_1 & \dots & b_N\end{bmatrix} = \begin{bmatrix} -2 & * & * & & \dots & * \\ * & \lambda_N - \lambda_1 & 0 & & \dots & 0 \\ * & 0 & \ddots & & & 0 \\ \vdots & \vdots & & \ddots & & \vdots \\ * & 0 & & & \lambda_3 - \lambda_1 & 0 \\ * & 0 & & \dots & 0 & \lambda_2 - \lambda_1 \end{bmatrix},$$

which is of the general form

$$J'(c^\star, u^\star) = \begin{bmatrix} E_J & S_J \\ -S_J^\top & C_J \end{bmatrix}$$

with $S_J = (2v_1^\top(G+G^\top)v_N, \ldots, 2v_1^\top(G+G^\top)v_2)$ and $E_J$ negative definite. The matrix $C_J$ is negative definite if the eigenvalue $\lambda_1(F(c^\star))$ is simple and negative semidefinite otherwise. In the case that $\lambda_1(F(c^\star))$ is simple, choosing $\mathcal{X} = I_N$ in

$$\begin{bmatrix} E_J & -S_J \\ S_J^\top & C_J \end{bmatrix} \mathcal{X} + \mathcal{X} \begin{bmatrix} E_J & S_J \\ -S_J^\top & C_J \end{bmatrix} = 2 \begin{bmatrix} E_J & 0 \\ 0 & C_J \end{bmatrix}$$

yields a negative-definite matrix. Applying Lyapunov's theory for linear systems and the Hartman-Grobman theorem, this proves local exponential stability of $(c^\star, u^\star)$ in the case that $\lambda_1(F(c^\star))$ is simple.

In the case that $\lambda_1(F(c^\star))$ has multiplicity of two, we continue by applying the *Ky Fan* inequality [12, Proposition III.5.3] to find that

$$\operatorname{Re}\left(\lambda_i(J'(c^\star, u^\star))\right) \le \lambda_1\left(\frac{1}{2}\left(J'(c^\star, u^\star) + J'(c^\star, u^\star)^\top\right)\right)$$

holds for all $i = 1, \ldots, N$. Since $\frac{1}{2}\left(J'(c^\star, u^\star) + J'(c^\star, u^\star)^\top\right)$ denotes the symmetric part of the Jacobian on the tangent space $T_{(c^\star, u^\star)}\left(\mathbb{R} \times \mathbb{S}^{N-1}\right)$, which is negative semidefinite, we know that $\operatorname{Re}\left(\lambda_i(J'(c^\star, u^\star))\right) \le 0$ for all $i = 1, \ldots, N$.

Furthermore, we need to exclude possible eigenvalues on the imaginary axis. In [73, Theorem 2], Ostrowski and Schneider draw a connection between purely imaginary eigenvalues of a matrix $J'(c^\star, u^\star)$ and conditions on its symmetric part $\frac{1}{2}\left(J'(c^\star, u^\star) + J'(c^\star, u^\star)^\top\right)$. More precisely, if $\frac{1}{2}\left(J'(c^\star, u^\star) + J'(c^\star, u^\star)^\top\right)$ is semidefinite and real, then the corresponding eigenvectors to $k = 2j$ imaginary eigenvalues $(\pm i\alpha_1, \ldots, \pm i\alpha_j)$ of $J'(c^\star, u^\star)$ are in the nullspace of $\frac{1}{2}\left(J'(c^\star, u^\star) + J'(c^\star, u^\star)^\top\right)$. If $\lambda_1(F(c^\star))$ has multiplicity of two, we find that the symmetric part of $J'(c^\star, u^\star)$, which reads

$$\frac{1}{2}\left(J'(c^\star, u^\star) + J'(c^\star, u^\star)^\top\right) = \operatorname{diag}\left(-2, \lambda_N - \lambda_1, \ldots, \lambda_3 - \lambda_1, 0\right), \tag{3.30}$$

has only a one dimensional nullspace. Since any eigenvalues on the imaginary axis would correspond to an even number of eigenvectors that must lie in the nullspace of (3.30), we can conclude that $J'(c^\star, u^\star)$ has no purely imaginary eigenvalues.

Finally, we need to investigate possible zero eigenvalues. Rearranging rows of $J'(c^\star, u^\star)$ yields

$$\begin{bmatrix} -a & 0 & 0 & \dots & 0 \\ * & (\lambda_N - \lambda_1) & 0 & \dots & 0 \\ \vdots & & \ddots & & \vdots \\ * & & & (\lambda_3 - \lambda_1) & 0 \\ -2 & * & * & \dots & a \end{bmatrix}$$

with $a = v_2^\top (G + G^\top) u^\star$, which is a triangular matrix with nonzero entries on the diagonal if and only if $v_2^\top (G + G^\top) u^\star \neq 0$. This reveals that $J'(c^\star, u^\star)$ has full rank and therefore no zero eigenvalue under Assumption 1.

In summary, the linearization of the dynamics in (3.29) on the manifold $\mathbb{R} \times \mathbb{S}^{N-1}$ at the equilibrium point $(c^\star, u^\star)$ leads to a Jacobian $J'(c^\star, u^\star)$ with $\mathrm{Re}\,(\lambda_i(J'(c^\star, u^\star))) < 0$. Hence, in the tangent space $T_{(c^\star, u^\star)}\left(\mathbb{R} \times \mathbb{S}^{N-1}\right)$, the point $(c^\star, u^\star)$ is locally exponentially stable. This concludes the proof. ■

Let us shortly discuss the case when the eigenvalue $\lambda_1(F(c^\star))$ is an eigenvalue of multiplicity two and Assumption 1 does not hold, i.e., $v_2^\top (G + G^\top) u^\star = 0$. This happens only in the technical case that at least one of the two analytic eigenvalue functions $\tilde{\lambda}_{i=1,2}(c)$, from rearrangement of $\lambda_{i=1,2}(F(c))$ [80], that meet at $(c^\star, \lambda_1(F(c^\star)))$, has a vanishing gradient at $c^\star$. We refer the interested reader to [59, 80] for more details on eigenvalues of Hermitian matrix functions.

**Remark 3.4.** As a corollary to Theorem 3.4, we can show that the optimizer $(c^\star, u^\star)$ is a local min-max saddle point of $\rho_3$ via the Taylor series expansion given by

$$\begin{aligned} \rho_3(c, u^\star) &= \rho_3(c^\star, u^\star) + \nabla_c \rho_3(c^\star, u^\star) \delta c + \frac{1}{2} \nabla_{cc} \rho_3(c^\star, u^\star) \delta c^2 + \mathcal{O}(\delta c^3), \\ \rho_3(c^\star, u) &= \rho_3(c^\star, u^\star) + \nabla_u \rho_3(c^\star, u^\star) \delta u + \frac{1}{2} \delta u^\top \nabla_{uu} \rho_3(c^\star, u^\star) \delta u + \mathcal{O}(\delta u^3). \end{aligned}$$

With $\nabla_u \rho_3(c^\star, u^\star) = \nabla_c \rho_3(c^\star, u^\star) = 0$, $\nabla_{uu} \rho_3(c^\star, u^\star)$ positive semidefinite, and $\nabla_{cc} \rho_3(c^\star, u^\star)$ negative definite on the tangent space $T_{(c^\star, u^\star)}\left(\mathbb{R} \times \mathbb{S}^{N-1}\right)$, the inequality $\rho_3(c^\star, u) \leq \rho_3(c^\star, u^\star) \leq \rho_3(c, u^\star)$ holds in a neighborhood of $(c^\star, u^\star)$, and thus $(c^\star, u^\star)$ is in fact a locally exponentially stable local min-max saddle point of $\rho_3$ on the manifold $\mathbb{R} \times \mathbb{S}^{N-1}$.

**Remark 3.5.** The structure of the Jacobian $J(c^\star, u^\star)$ relates the linearization of (3.29) around the critical point $(c^\star, u^\star)$ to the well-known (linear) saddle point problems that arise, for example, in the context of regularized weighted least-squares problems, from certain interior point methods in optimization, or from Lagrange functions [7].

**Conic relations - discrete-time solution**

Iterative approaches for the solution of saddle point problems have been introduced, for example, in the book of Arrow, Hurwicz, and Uzawa [3] and in an article of Polyak [77]. In these references, iterative schemes consisting of simultaneous iterations in both variables and their convergence are discussed, addressing mainly the problem of finding the saddle point of a Lagrangian. One of the iterative approaches introduced in [3, Chapter 10, Sections 4-5], is the so-called *Arrow-Hurwicz* iteration, which reads

$$\begin{aligned} c^{(i+1)} &= c^{(i)} - \alpha \nabla_c \rho_3(c^{(i)}, u^{(i)}), \\ u^{(i+1)} &= u^{(i)} + \alpha \nabla_u \rho_3(c^{(i)}, u^{(i)}). \end{aligned} \tag{3.31}$$

In [AK19], it is shown along the lines of [77] that for a small enough step size $\alpha$, the method stated in (3.31) is locally convergent to $(c^\star, u^\star)$, and the modified *Arrow-Hurwicz* method [78] is introduced as an expedient method to determine the minimal cone of an unknown input-output system.

The *Uzawa* iteration for general saddle point problems, also called the dual method, was presented by Uzawa in [3, Chapter 10]. Here, the gradient iteration is only performed with respect to the input $u$, while the corresponding center $c^{(i)}$ is found by minimization of $\rho_3(c, u^{(i)})$ with respect to $c$:

$$\begin{aligned} c^{(i+1)} &= \arg\min_{c \in \mathbb{R}} \rho_3(c, u^{(i)}), \\ u^{(i+1)} &= u^{(i)} + \alpha \nabla_u \rho_3(c^{(i)}, u^{(i)}). \end{aligned} \tag{3.32}$$

For any given $u^{(i)} \in \mathbb{S}^{N-1}$, $\rho_3(\cdot, u^{(i)})$ is a strongly convex function with a global minimum at the critical point $c = 0.5(u^{(i)\top}(G + G^\top)u^{(i)})$ (cf. Lemma 3.2). With the results from Proposition 3.5, we can hence compute $\min_{c\in\mathbb{R}} \rho_3(c, u^{(i)}) = 0.5u^{(i)\top}(T_P G T_P u^{(i)} + G u^{(i)})$ with two input-output tuples and $\nabla_u \rho_3(c^{(i)}, u^{(i)})$ with one additional input-output tuple from (numerical) experiments.

Hence, the *Uzawa* iteration on the manifold $\mathbb{R} \times \mathbb{S}^{N-1}$ is given by

$$\begin{aligned} c_{k+1} &= 0.5u^{(i)\top}(T_P G T_P u^{(i)} + Gu^{(i)}), \\ \tilde{u}^{(i+1)} &= u^{(i)} + 2\alpha(T_P G T_P G u^{(i)} - c(Gu^{(i)} + T_P G T_P u^{(i)})) \\ &\quad - 2\alpha u^{(i)\top}(T_P G T_P G u^{(i)} - c(Gu^{(i)} + T_P G T_P u^{(i)}))u^{(i)}, \\ u^{(i+1)} &= \frac{\tilde{u}^{(i+1)}}{\|\tilde{u}^{(i+1)}\|}, \end{aligned} \tag{3.33}$$

with step size $\alpha$, where we applied again a valid retraction mapping from the tangent space $T_{(c^{(i)},u^{(i)})}(\mathbb{R} \times \mathbb{S}^{N-1})$ to the manifold $\mathbb{R} \times \mathbb{S}^{N-1}$ [1]. Along the lines of [77], we show in the following that (3.33) is locally convergent to $(c^\star, u^\star)$.

**Proposition 3.6.** *Assume that $\lambda_1(F(c^\star))$ is an eigenvalue*

- *with multiplicity one, or*
- *with multiplicity two and Assumption 3.1 holds.*

*Then, there exists an $\bar{\alpha}$ such that for all $\alpha \in (0, \bar{\alpha})$ the iteration described in (3.33) is locally convergent to the equilibrium point $(c^\star, u^\star)$ with $r^2_{N,\min} = \rho_3(c^\star, u^\star)$.*

*Proof.* The local behavior of the *Uzawa* iteration is given by

$$e^{(i+1)} = K(c^\star, u^\star)e^{(i)}, \quad e^{(i)} = \begin{bmatrix} c^{(i)} \\ u^{(i)} \end{bmatrix} - \begin{bmatrix} c^\star \\ u^\star \end{bmatrix},$$

where

$$K(c^\star, u^\star) = \begin{bmatrix} 0 & -\frac{1}{2}\nabla_{cu}\rho_3(c^\star, u^\star) \\ \alpha\nabla_{uc}\rho_3(c^\star, u^\star) & I_n + \alpha\nabla_{uu}\rho_3(c^\star, u^\star) \end{bmatrix}.$$

To improve readability, we denote $\nabla_{uc}\rho_3(c^\star, u^\star)$ by $S_\rho$ in the following.

Since the projection onto the manifold $\mathbb{R} \times \mathbb{S}^{N-1}$ is a smooth retraction mapping from any tangent space onto the manifold, this projection preserves convergence properties of the algorithm [1, Chapter 4]. Therefore, we are interested in the eigenvalues of $K(c^\star, u^\star)$ on the tangent space $T_{(c^\star,u^\star)}(\mathbb{R} \times \mathbb{S}^{N-1})$ spanned by the vectors $b_1, \ldots, b_N$.

The eigenvalue equation for the matrix $K(c^\star, u^\star)$ with the eigenvalue $\mu$ and the eigenvector $(c_e, u_e)$ reads

$$\begin{aligned} -\frac{1}{2} S_\rho^\top u_e &= \mu c_e, \\ \alpha S_\rho c_e + (I_N + \alpha \nabla_{uu} \rho_3(c^\star, u^\star)) u_e &= \mu u_e, \\ u_e^\top u^\star &= 0. \end{aligned}$$

Multiplying the first equation by $\alpha S_\rho$ and replacing $\alpha S_\rho c_e$ by the second equation yields

$$-\frac{\alpha}{2} S_\rho S_\rho^\top u_e = \mu((\mu - 1) I_N - \alpha \nabla_{uu} \rho_3(c^\star, u^\star)) u_e,$$

and thus

$$\frac{\alpha}{2} \left( S_\rho S_\rho^\top - 2 \nabla_{uu} \rho_3(c^\star, u^\star) \right) u_e = \mu(1 - \mu) u_e.$$

If $u_e = 0$, then $\mu c_e = 0$. Since $(c_e, u_e) \neq 0$, and thus $c_e \neq 0$, we find $\mu = 0$ implying $|\mu| < 1$.

If $u_e \neq 0$, then $\mu(1-\mu)$ is an eigenvalue of the matrix $\frac{\alpha}{2} \left( S_\rho S_\rho^\top - 2 \nabla_{uu} \rho_3(c^\star, u^\star) \right)$, which is symmetric and positive definite in the tangent space $T_{(c^\star, u^\star)}(\mathbb{R} \times \mathbb{S}^{N-1})$ if $\lambda_1(F(c^\star))$ is an eigenvalue of multiplicity one or an eigenvalue of multiplicity two and Assumption 1 holds (cf. Theorem 3). Hence, we know that $\mu(1-\mu)$ is real and

$$0 < \mu(1 - \mu) \leq \alpha \left\| \frac{1}{2} S_\rho S_\rho^\top - \nabla_{uu} \rho_3(c^\star, u^\star) \right\|. \tag{3.34}$$

The term $\mu(1-\mu)$ being real implies $\mathrm{Im}(\mu) = 0$, or $\mathrm{Re}(\mu) = \frac{1}{2}$. With $\mathrm{Im}(\mu) = 0$ and $\mu(1-\mu) > 0$, it follows directly that $|\mu| < 1$ must hold.

Let

$$\bar{\alpha} = \frac{1}{2 \| S_\rho S_\rho^\top - 2 \nabla_{uu} \rho_3(c^\star, u^\star) \|}.$$

In the case $\mathrm{Im}(\mu) \neq 0$, and hence $\mathrm{Re}(\mu) = \frac{1}{2}$, we find $\mu(1-\mu) = \frac{1}{4} + \mathrm{Im}^2(\mu) > \frac{1}{4} = \bar{\alpha} \left\| \frac{1}{2} S_\rho S_\rho^\top - \nabla_{uu} \rho_3(c^\star, u^\star) \right\|$. This, however, contradicts (3.34) for all $\alpha < \bar{\alpha}$, and thus we have $\mathrm{Im}(\mu) = 0$.

Altogether, this leaves us with eigenvalues $\mu$ with an absolute value strictly less than one whenever $\alpha < \bar{\alpha}$, and hence $(e^{(i)}) \to 0$ for $i \to \infty$ whenever $e^{(0)}$ is small. Therefore, the iteration described in (3.33) is locally convergent to $(c^\star, u^\star)$. ∎

With this iterative approach for saddle point problems, we conclude this section of methods for SISO discrete-time LTI systems to identify the $\mathcal{L}_2$-gain, passivity properties, and the minimal cone that an a priori unknown system is confined to. The presented analysis for sampling-based inference of control-theoretic input-output properties depicts the possibilities of this framework and builds the basis for extensions, improvements, or its application to other classes of systems.

## 3.2 Generalizations and extensions

In the previous section, we introduced a systematic approach to iteratively determine certain dissipativity properties from input-output data and provided a rigorous mathematical framework, which builds the foundation for generalizations and extensions. In this section, we start by introducing the necessary tools to evaluate also continuous-time systems via iterative methods and show how a similar approach than in the previous section is applicable. Furthermore, we summarize how the presented results can be applied to MIMO systems, discuss how measurement noise impacts the approach, and present some insights into the convergence rate.

### 3.2.1 Continuous-time LTI systems

In this section, we consider SISO continuous-time LTI systems $H : l_{2,\mathrm{c}} \to l_{2,\mathrm{c}}$. The input to output operator $H$ can be written as

$$y(t) = (g * u)(t) = \int g(t-\zeta)u(\zeta)\,\mathrm{d}\zeta$$

where $g$ denotes the continuous-time impulse response of the system. In the following, the convolution operator $u \mapsto g * u$ will be denoted by $u \mapsto C_g(u)$ for readability. Again, we assume $u(t) = 0$ for $t < 0$.

For every bounded linear operator on a Hilbert space $H : l_{2,\mathrm{c}} \to l_{2,\mathrm{c}}$, there exists a unique adjoint operator $H^\star : l_{2,\mathrm{c}} \to l_{2,\mathrm{c}}$ defined by $\langle H(u), y\rangle = \langle u, H^\star(y)\rangle$, wherein $\langle\cdot,\cdot\rangle : l_{2,\mathrm{c}} \times l_{2,\mathrm{c}} \to \mathbb{R}$ denotes the $l_{2,\mathrm{c}}$-inner product and $\|\cdot\| : l_{2,\mathrm{c}} \to \mathbb{R}$ denotes the

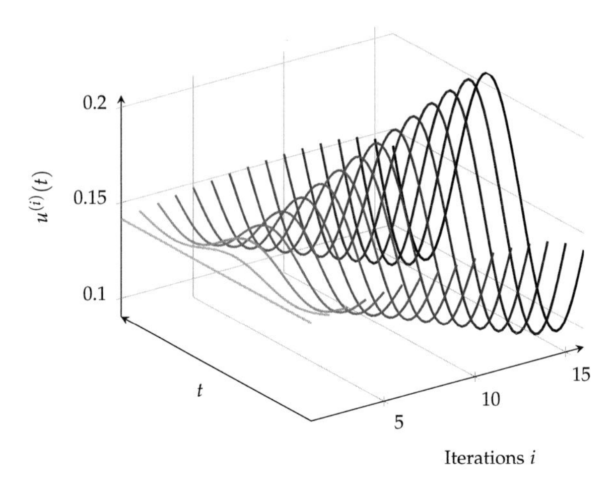


**Figure 3.9.** The goal is to iteratively converge from an initial input $u^{(0)} \in l_{2,c}$ towards the input $u$ corresponding to the system property of interest, e.g., the $\mathcal{L}_2$-gain $\gamma$.

$l_{2,c}$-norm. Let $\bar{g}(t) = g(-t)$. Then

$$\begin{aligned} \langle C_g(u), y\rangle &= \int \int g(t-\zeta)u(\zeta)\,\mathrm{d}\zeta\, y(t)\,\mathrm{d}t = \int \int g(t-\zeta)y(t)u(\zeta)\,\mathrm{d}t\,\mathrm{d}\zeta \\ &= \int u(\zeta) \int g(-(\zeta - t))y(t)\,\mathrm{d}t\,\mathrm{d}\zeta = \langle u, C_{\bar{g}}(y)\rangle \end{aligned} \tag{3.35}$$

verifies that $C_{\bar{g}}$ is the adjoint operator of $C_g$.

In the following, we iteratively search for the input $u \in l_{2,c}$ corresponding to the $\mathcal{L}_2$-gain $\gamma$ (or the input feedforward passivity parameter $\nu$, the shortage of passivity $\beta$, or conic relations $(c, r)$, respectively), as depicted in Fig. 3.9. For continuous-time LTI systems, we can reformulate the $\mathcal{L}_2$-gain condition in (3.3) into

$$\begin{aligned} \gamma^2 &= \sup_{u \in l_{2,c},\, \|u\| \neq 0} \rho_1(u) = \sup_{u \in l_{2,c},\, \|u\| \neq 0} \frac{1}{2} \frac{\langle C_g(u), C_g(u)\rangle}{\|u\|^2} \\ &= \sup_{u \in l_{2,c},\, \|u\| \neq 0} \frac{\int \left(\int g(\zeta)u(t-\zeta)\,\mathrm{d}\zeta\right)^2 \mathrm{d}t}{\int u^2(t)\,\mathrm{d}t} \end{aligned}$$

where $\rho_1 : l_{2,c} \setminus \{0\} \to \mathbb{R}$ is a scale-invariant function, also referred to as the Rayleigh quotient. Without loss of generality, we consider $\rho_1$ on the unit sphere $S_{l_{2,c}} = \{u \in$

$l_{2,c} | \|u\| = 1\}$. We say that $\rho_1 : S_{l_{2,c}} \to \mathbb{R}$ is Fréchet-differentiable if for every $u \in S_{l_{2,c}}$ there exists a linear operator $D\rho_1(u) : l_{2,c} \to \mathbb{R}$ such that

$$\lim_{h \to 0} \frac{|\rho_1(u+h) - \rho_1(u) - D\rho_1(u)(h)|}{\|h\|} = 0.$$

By the Riesz representation theorem, there is a unique $\rho_1' : l_{2,c} \to l_{2,c}$ such that $D\rho_1(u)(h) = \langle \rho_1'(u), h \rangle$ if $\rho_1$ is Fréchet-differentiable at $u$ [90]. $D\rho_1(u)$ is also called the dual of $\rho_1'(u)$.

**Proposition 3.7.** *The Fréchet-derivative of $\rho_1$ on the unit sphere $S_{l_{2,c}}$ is given by*

$$\rho_1'(u) = 2\left(C_{\bar{g}} \circ C_g\right)(u) - 2\rho_1(u)u$$

*and can be computed by evaluating $u \mapsto C_g(u)$ twice.*

*Proof.* First, we claim that the Fréchet-derivative of $f(u) = \langle C_g(u), C_g(u) \rangle$ is given by $f'(u) = 2\left(C_{\bar{g}} \circ C_g\right)(u)$. By applying the presented definition, knowledge of the adjoint operator of $C_g$ as given in (3.35), and the linearity of $C_g$, we find

$$\begin{aligned}
&|f(u+h) - f(u) - \langle f'(u), h \rangle| \\
=&|\langle C_g(u+h), C_g(u+h) \rangle - \langle C_g(u), C_g(u) \rangle - 2\langle \left(C_{\bar{g}} \circ C_g\right)(u), h \rangle| \\
=&|\langle C_g(h), C_g(h) \rangle| = \mathcal{O}(\|h\|^2),
\end{aligned}$$

and, thus, that the claim is indeed true. From the quotient rule follows

$$\rho_1'(u) = \frac{2}{\|u\|^2}\left(\left(C_{\bar{g}} \circ C_g\right)(u) - \frac{\langle C_g(u), C_g(u) \rangle}{\langle u, u \rangle} u\right) = 2\left(C_{\bar{g}} \circ C_g\right)(u) - 2\rho_1(u)u.$$

In order to calculate the Fréchet-derivative $\rho_1'(u)$ from two input-output tuples, we require $C_g(u)$ and $C_{\bar{g}} \circ C_g(u)$ from evaluating $u \mapsto C_g(u)$. Therefore, we rewrite the adjoint operator

$$C_{\bar{g}}(y)(t) = \int g(-\zeta)y(t-\zeta)\,\mathrm{d}\zeta = \int g(\zeta)y(t+\zeta)\,\mathrm{d}\zeta = \int g(\zeta)\bar{y}(-t-\zeta)\,\mathrm{d}\zeta = C_g(\bar{y})(-t)$$

to see that $\left(C_{\bar{g}} \circ C_g\right)(u) = C_{\bar{g}}(y) = \overline{C_g(\bar{y})} = \overline{C_g(\overline{C_g(u)})}$ holds, where the bar again denotes time-reversal. ∎

Hence, even though we have no knowledge about the system but input-output information, we can construct the Fréchet-derivative $\rho_1'(u)$ from two input-output samples. Analogously to the discrete-time case, these two input-output samples are $C_g(u)$ and $(C_{\bar{g}} \circ C_g)(u) = \overline{C_g(\overline{C_g(u)})}$, which is obtained by time-reversing the output $\overline{C_g(u)}(t) = C_g(u)(-t)$, choosing it as yet another input, and time-reversing the output again.

Similar to the discrete-time case, we plan to maximize $\rho_1$ by a dynamical system that is now described by the evolution equation

$$\frac{\partial}{\partial \tau} u(\tau) = 2\left(C_{\bar{g}} \circ C_g\right)(u(\tau)) - 2\rho_1(u(\tau))u(\tau) \tag{3.36}$$

along whose solution $\rho_1$ increases monotonically. We can show that (3.36) leaves the unit sphere $S_{l_{2,c}}$ invariant:

$$\frac{\partial}{\partial \tau}\|u(\tau)\|^2 = 2\langle u(\tau), \frac{\partial}{\partial \tau} u(\tau)\rangle = 4\langle u(\tau), \left(C_{\bar{g}} \circ C_g\right)(u(\tau)) - \rho_1(u(\tau))u(\tau)\rangle = 0.$$

**Proposition 3.8.** *The Rayleigh quotient $\rho_1(u(\tau))$ monotonically increases along the solutions of* (3.36) *and converges for $\tau \to \infty$.*

*Proof.* According to the Courant-Fischer-Weyl principle for self-adjoint operators we find an upper bound on $\rho_1$ by $\sup_{u \in l_{2,c},\, \|u\| \neq 0} \rho_1(u) = \sup \sigma$, where $\sigma$ denotes the spectrum of the linear, self-adjoint, and bounded operator $u \mapsto C_{\bar{g}} \circ C_g(u)$. This principle is also referred to as the Rayleigh-Ritz principle. Moreover, on the basis of the Fréchet-derivative of Proposition 3.7, we can conclude that $\rho_1(u(\tau))$ is monotonically increasing along the solutions of (3.36):

$$\frac{\partial}{\partial \tau}\rho_1(u(\tau)) = \langle \rho_1'(u(\tau)), \frac{\partial}{\partial \tau} u(\tau)\rangle = \|\rho_1'(u(\tau))\|^2 \geq 0. \tag{3.37}$$

Thus, $\rho_1(u(\tau))$ is monotonically increasing with $\tau$ and upper-bounded by the Rayleigh-Ritz principle stated above. By the monotone convergence theorem, $\tau \mapsto \rho_1(u(\tau))$ converges. ∎

The above results show how the general concept of iterative sampling strategies to infer input-output properties can also be extended to continuous-time systems. In practice, the fact that only experiments of finite length can be performed need again to be taken into account. Analogously to the case of the gain as introduced above, also the gradients of the optimization problem with respect to passivity or conic relations can be obtained from evaluating $u \mapsto C_g(u)$ (cf. [AK8]).

### 3.2.2 MIMO systems

In this subsection, we shortly summarize how the presented approach can be extended to MIMO systems, which has been introduced in the special case of the $\mathcal{L}_2$-gain in [72]. For simplicity, we consider square MIMO systems ($p = m$), for which the input-output map for a given finite-length input sequence in matrix notation reads

$$\begin{bmatrix} y^{[1]} \\ y^{[2]} \\ \vdots \\ y^{[m]} \end{bmatrix} = \begin{bmatrix} G_{11} & G_{12} & \cdots & G_{1m} \\ G_{21} & G_{22} & \cdots & G_{2m} \\ \vdots & \vdots & \vdots & \vdots \\ G_{m1} & G_{m2} & \cdots & G_{mm} \end{bmatrix} \begin{bmatrix} u^{[1]} \\ u^{[2]} \\ \vdots \\ u^{[m]} \end{bmatrix} \tag{3.38}$$

with $u^{[j]} = \begin{bmatrix} u_0^{[j]} & \cdots & u_{N-1}^{[j]} \end{bmatrix}^\top$, $y^{[j]} = \begin{bmatrix} y_0^{[j]} & \cdots & y_{N-1}^{[j]} \end{bmatrix}^\top$, $u^{[j]}, y^{[j]} \in \mathbb{R}^N$, $j = 1, \ldots, m$. In short notation, (3.38) will be denoted by $\mathtt{Y} = \mathtt{GU}$ where $\mathtt{Y} \in \mathbb{R}^{mN}$, $\mathtt{U} \in \mathbb{R}^{mN}$, and $\mathtt{G} \in \mathbb{R}^{mN \times mN}$. Note that in the case of output strict passivity, we assume $\mathtt{G}$ to have full rank, similarly to the SISO consideration. The $\mathcal{L}_2$-gain, input strict as well as output strict passivity can then again be formulated in terms of optimization problems

$$\gamma_N^2 = \max_{\mathtt{U} \in \mathbb{R}^{mN}, \|\mathtt{U}\| \neq 0} \frac{\|\mathtt{GU}\|^2}{\|\mathtt{U}\|^2}, \quad \nu_N = \min_{\mathtt{U} \in \mathbb{R}^{mN}, \|\mathtt{U}\| \neq 0} \frac{\mathtt{U}^\top \mathtt{GU}}{\|\mathtt{U}\|^2}, \quad \beta_N = - \min_{\mathtt{U} \in \mathbb{R}^{mN}, \|\mathtt{U}\| \neq 0} \frac{\mathtt{U}^\top \mathtt{GU}}{\|\mathtt{GU}\|^2}.$$

For these input-output system properties, we hence retrieve the same optimization problems as in the SISO case with the same respective gradients in (3.5), (3.15), and (3.19), which can be computed with the terms $\mathtt{U}, \mathtt{GU}, \mathtt{G}^\top \mathtt{U}$, and $\mathtt{G}^\top \mathtt{GU}$. This also holds in the case of conic relations when parameterizing the center matrix by $cI_m$ yielding the optimization problem

$$r_{N,\min}^2 = \min_{c \in \mathbb{R}} \max_{\mathtt{U} \in \mathbb{R}^{mN}, \|\mathtt{U}\| \neq 0} \frac{c^2 \|\mathtt{U}\|^2 - 2c\mathtt{U}^\top \mathtt{GU} + \|\mathtt{GU}\|^2}{\|\mathtt{U}\|^2},$$

with the resulting gradients in (3.27) and (3.28).

In contrast to the SISO case, however, $\mathtt{G}$ does not have Toeplitz structure, and therefore $T_\mathrm{P}\mathtt{G}^\top \neq \mathtt{G}T_\mathrm{P}$. Nevertheless, we can still compute the gradients by evaluating $\mathtt{U} \mapsto \mathtt{GU}$ while $\mathtt{G}$ remains undisclosed. Let $E_m^{ij}$ be a $m \times m$ matrix with zero entries everywhere except for the single entry 1 at the $i$th row and $j$th column. Decomposing $\mathtt{G}^\top \mathtt{U} = \sum_{i,j=1}^m (E_m^{ij} \otimes T_\mathrm{P}) \mathtt{G} (E_m^{ij} \otimes T_\mathrm{P}) \mathtt{U}$ yields a constructive procedure to compute $\mathtt{G}^\top \mathtt{U}$ from $m^2$ input-output tuples. For all $i, j = 1, \ldots, m$, we choose the $j^{th}$ component of

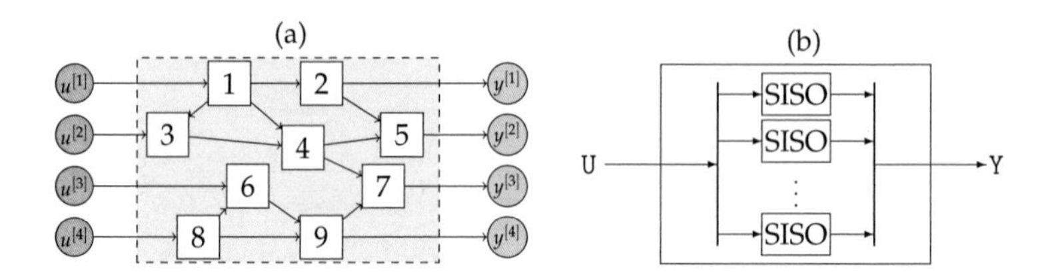


**Figure 3.10.** Examples of MIMO systems with known couplings such as a networked dynamical systems with known interconnections in (a) or a composite system consisting of subsystems operated in parallel in (b).

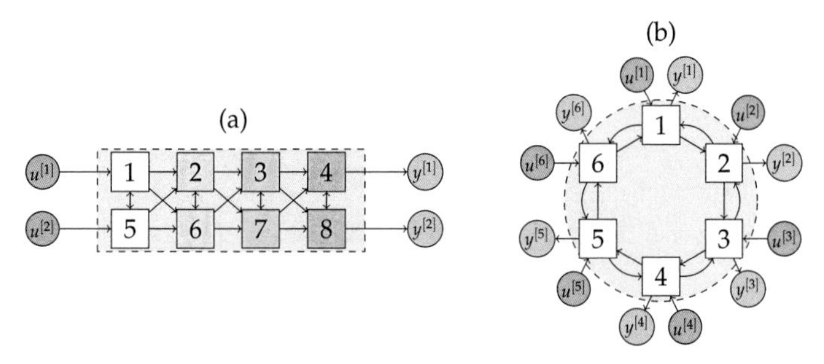


**Figure 3.11.** Examples of MIMO systems consisting of symmetrically interconnected subsystems such as a two-lane vehicle platoon in (a) or a cyclic network in (b).

$u$, viz. $u^{[j]}$, time-reverse it, apply it to the $i^{th}$ input of the system, measure only the $j^{th}$ output, and time reverse it again. We hence require $m^2 + 1$ evaluations of $\mathtt{U} \mapsto \mathtt{GU}$ to compute $\nabla \rho_1(\mathtt{U})$, and $2m^2 + 1$ evaluations to calculate $\nabla \rho_2(\mathtt{U})$, or $\nabla_c \rho_3(c, \mathtt{U})$ together with $\nabla_{\mathtt{U}} \rho_3(c, \mathtt{U})$. However, prior knowledge on the coupling of MIMO systems can reduce the number of required experiments per iteration significantly. Such prior knowledge reducing the number of required data tuples includes, for example, knowledge on the interconnection of networked dynamical systems, as illustrated in Figure 3.10(a), or its special case of composite systems, as depicted in Figure 3.10(b). Moreover, knowledge on symmetries in the coupling can also be particularly valuable in this context, and special symmetries, as exemplarily depicted in Figure 3.11, even allow to simply apply the sampling scheme for SISO systems. The reader is referred to [AK22] for more details and the respective proofs.

Since the optimization problems are analogous to the SISO case, and since we have shown that the gradient can be computed from (numerical) experiments, all convergence guarantees presented in Section 3.1 also hold for MIMO systems and can be extended along the lines of Section 3.2.1 and Section 3.2.3.

### 3.2.3 Measurement noise

The presented framework for determining system properties is based on gradient dynamical systems. Generally speaking, the iterative procedure hence inherits robustness properties of such approaches from classical results, e.g., from [76]. To be more specific, we evaluate the case where the output is corrupted by additive measurement noise $e = \begin{bmatrix} e_0 & \cdots & e_{N-1} \end{bmatrix}^\top$. Similar to [106], we consider white noise with zero mean and variance $\sigma_e^2$. For the $\mathcal{L}_2$-gain, this implies that the data tuples of the experiments necessary for calculating the gradient read $(u^{(i)}, Gu^{(i)}+e^{(i),1})$ and $(T_P(Gu^{(i)}+e^{(i),1}), GT_PGu^{(i)}+GT_Pe^{(i),1}+e^{(i),2})$, where $e^{(i),1}$ and $e^{(i),2}$ are the measurement noise sequences of the first and second experiment of iteration $i$, respectively.

**Lemma 3.3.** *The gradient $\hat{\nabla}\rho_1(u^{(i)})$ computed via Proposition 3.1 by evaluating $u^{(i)} \mapsto Gu^{(i)} + e^{(i),j}$ twice ($j = 1, 2$), with $e^{(i),j} = \begin{bmatrix} e_0^{(i),j}, \dots, e_{N-1}^{(i),j} \end{bmatrix}^\top$, $e_k^{(i),j} \sim \mathcal{N}(0, \sigma_e^2)$, $k = 0, \dots, N-1$, yields*

$$\hat{\nabla}\rho_1(u^{(i)}) = \nabla\rho_1(u^{(i)}) + \epsilon^{(i)} \tag{3.39}$$

*with $\mathbb{E}[\epsilon^{(i)}] = 0$ and $\mathbb{E}[\|\epsilon^{(i)}\|^2] \leq 4\mathbb{E}[e^{(i),1\top}GG^\top e^{(i),1} + e^{(i),2\top}e^{(i),2}]$.*

*Proof.* Computing the gradient vector field $\rho_1 : \mathbb{S}^{N-1} \to \mathbb{R}$ from the noise corrupted data $T_P(Gu^{(i)} + e^{(i),1}) \mapsto GT_PGu^{(i)} + GT_Pe^{(i),1} + e^{(i),2}$ yields (3.39) with

$$\epsilon^{(i)} = 2\left(T_PGT_Pe^{(i),1} + T_Pe^{(i),2} - \left(u^{(i)\top}T_PGT_Pe^{(i),1} + u^{(i)\top}T_Pe^{(i),2}\right)u^{(i)}\right).$$

The linearity of the expectation operator and $e_k^{(i),1}, e_k^{(i),2} \sim \mathcal{N}(0, \sigma_e^2)$, $k = 0, \dots, N-1$, directly leads to $\mathbb{E}[\epsilon_k^{(i)}] = 0$, $k = 0, \dots, N-1$ as well as to an upper bound on the variance

$$\begin{aligned}\mathbb{E}[\|\epsilon^{(i)}\|^2] &= 4\mathbb{E}[e^{(i),1\top}GG^\top e^{(i),1} + e^{(i),2\top}e^{(i),2} - (u^{(i)\top}G^\top e^{(i),1})^2 - (u^{(i)\top}T_Pe^{(i),2})^2] \\ &\leq 4\mathbb{E}[e^{(i),1\top}GG^\top e^{(i),1} + e^{(i),2\top}e^{(i),2}].\end{aligned}$$

■

Most importantly, this means that the gradient is unbiased. The upper bound on the variance provides theoretical insights, while its exact calculation would require prior knowledge of some bound on G. Similar results hold for the gradient of the input strict passivity cost function in (3.19) and the gradients in (3.27)-(3.28) for conic relations. Exemplarily, we keep considering the $\mathcal{L}_2$-gain and let $\lambda_1 \geq \cdots \geq \lambda_N$ denote the eigenvalues of $G^\top G$. In the following lemma, we establish some characteristics of the cost function $\rho_1$, which, together with the results from Lemma 3.3, allow to apply well-known robustness results from the literature.

**Lemma 3.4.** *The Rayleigh quotient $\rho_1 : \mathbb{S}^{N-1} \to R$ has the following characteristics:*

- *$\rho_1$ is a smooth function,*
- *$\rho_1$ is locally strongly concave at $v_1$ with the concavity parameter $l_\rho = \lambda_1 - \lambda_2 > 0$ if and only if the largest eigenvalue $\lambda_1$ is simple,*
- *$\nabla\rho_1$ is locally Lipschitz at $v_1$ with the Lipschitz constant $L_\rho = \lambda_1 - \lambda_N$.*

*Proof.* The Rayleigh quotient $\rho_1 : R^N \setminus \{0\} \to R$ is a smooth function [32], and hence $\rho_1 : \mathbb{S}^{N-1} \to R$ is smooth as well. Since the function $\rho_1$ is twice continuously differentiable, we can evaluate the Hessian of $\rho_1$ in the tangent space $T_{v_1}\mathbb{S}^{N-1}$ to determine local strong concavity and the corresponding concavity parameter $l_\rho$. The computation of the Hessian reveals $\mathtt{H}_{\rho_1}(v_1) = 2(G^\top G - \lambda_1 I_N)$. By projection onto the tangent space $T_{v_1}\mathbb{S}^{N-1}$, which is spanned by the orthonormal vectors $v_2, \ldots, v_N$, we find

$$\begin{bmatrix} v_2 & \ldots & v_N \end{bmatrix}^\top \mathtt{H}_{\rho_1}(v_1) \begin{bmatrix} v_2 & \ldots & v_N \end{bmatrix} = 2\mathrm{diag}\left((\lambda_2 - \lambda_1), \ldots, (\lambda_N - \lambda_1)\right) \preceq (\lambda_2 - \lambda_1) I_{N-1},$$

and thus that $\rho_1$ is indeed locally strongly concave at $v_1$ on the manifold $\mathbb{S}^{N-1}$ with the concavity parameter $l_\rho = \lambda_1 - \lambda_2$.

Since $\rho_1$ is twice differentiable and locally concave at $v_1$ on the unit sphere $\mathbb{S}^{N-1}$, $\rho_1$ is locally Lipschitz with constant $L_\rho$ if and only if $\mathtt{H}_{\rho_1}(v_1) \succeq -L_\rho I_N$. The results above then finally lead to $L_\rho = \lambda_1 - \lambda_N$, which concludes the proof. ∎

Similar statements follow from Theorem 3.2 and Theorem 3.4 for $\rho_2$ and $\rho_3$. Lemma 3.3 together with Lemma 3.4 allow to directly apply results from [76], which we state here for general gradient methods in the presence of noise.

**Proposition 3.9.** *([76, Chapter 4, Theorem 3]) Let $f_\rho(u)$ be strongly concave (with constant $l_\rho$) with a gradient satisfying a Lipschitz condition (with constant $L_\rho$). Furthermore, let $u^{(i+1)} = u^{(i)} + \alpha^{(i)}(\nabla f_\rho(u^{(i)}) + \epsilon^{(i)})$ be the updating scheme where the noise $\epsilon^{(i)}$ is random, independent, with $\mathbb{E}[\epsilon^{(i)}] = 0$ and $\mathbb{E}[\|\epsilon^{(i)}\|^2] \leq \sigma^2$.*

*(i) Then there exists a $\bar{\alpha} > 0$ such that for $\alpha^{(i)} = \alpha$, $i = 1, 2, \ldots$, with $0 < \alpha < \bar{\alpha}$, we have $\mathbb{E}[f_\rho(u^\star) - f_\rho(u^{(i)})] \leq R(\alpha) + \mathbb{E}[f_\rho(u^\star) - f_\rho(u^{(0)})]q^i$ where $q < 1$, $R(\alpha) \to 0$ as $\alpha \to 0$.*

*(ii) If $\alpha^{(i)} \to 0$ and $\sum_{i=0}^{\infty} \alpha^{(i)} = \infty$, then $\mathbb{E}[\|u^{(i)} - u^\star\|^2] \to 0$.*

*(iii) Finally, if $\alpha^{(i)} = \frac{\alpha}{i}$ and $\alpha > \frac{1}{2l_\rho}$, then $\mathbb{E}[f_\rho(u^\star) - f_\rho(u^{(i)})] \leq \frac{L_\rho \sigma^2 \alpha^2}{2(2l_\rho\alpha - 1)i} + \mathcal{O}\left(\frac{1}{i}\right)$.*

With a suitably chosen step size $\alpha^{(i)}$, the iteration

$$\tilde{u}^{(i+1)} = u^{(i)} + \alpha^{(i)} \left(\nabla \rho_1(u^{(i)}) + \epsilon^{(i)}\right), \quad u^{(i+1)} = \frac{\tilde{u}^{(i+1)}}{\|\tilde{u}^{(i+1)}\|},$$

is hence locally convergent in mean square to $u^\star$ with $\rho_1(u^\star) = \gamma_N^2$, and similar results hold, e.g., for the input feedforward passivity index.

**Remark 3.6.** Even if the measurement noise cannot be characterized by zero mean white noise, but we can find a deterministic worst-case bound on $\epsilon^{(i)}$ in (3.39), i.e., $\epsilon^{(i)} \leq \bar{\epsilon}$, Theorem 1 in [76, Chapter 4] provides convergence guarantees for general gradient methods towards a neighborhood of the optimizer, the size of which is dependent on the noise level $\bar{\epsilon}$.

The above analysis also provides an approach to determine local convergence rates. Applying [44, Theorem 1.5] for a fixed step size of $\alpha = \frac{2}{L_\rho + l_\rho}$ leads to a local convergence estimate of

$$\rho_1(u^\star) - \rho_1(u^{(i)}) \leq \frac{L_\rho}{2} \left(\frac{L_\rho - l_\rho}{L\rho + l_\rho}\right)^{2i} \|u^\star - u^{(0)}\|^2.$$

For the gradient method with exact line search, which is possible without additional input-output tuples in the $\mathcal{L}_2$-gain and the input strict and output strict passivity case, we can apply [44, Theorem 1.2] to find the local convergence estimate of

$$\rho_1(u^\star) - \rho_1(u^{(i)}) \leq \left(\frac{L_\rho - l_\rho}{L_\rho + l_\rho}\right)^{2i} \left(\rho_1(u^{(0)}) - \rho_1(u^\star)\right). \tag{3.40}$$

More recently, [60] introduced design tools to tailor a gradient dynamical system to the required convergence rate and robustness (i.e., in [60], $H_2$-performance from noise to output/optimizer). Based on the results in Section 3.1, one can hence design an iterative gradient scheme with specific local robustness and convergence guarantees, e.g., for determining the $\mathcal{L}_2$-gain. This framework even paves the way towards extending the presented approaches to (slightly) nonlinear systems if the deviation between the gradient obtained from data and the true gradient of the cost function of interest can be bounded.

## 3.3 Numerical examples

In this section, we illustrate the applicability and the potential of the proposed methods with different examples, including an oscillator and a high-dimensional system.

### 3.3.1 $\mathcal{L}_2$-gain and conic relations of a random system

We start with a randomly generated LTI system of order 20 (MATLAB function *drss* with *rng(0)*), which has an $\mathcal{L}_2$-gain of $\gamma_{\text{tr}} = 13.7$. The initial input $u^{(0)} \in \mathbb{R}^{10^3}$ is $u_k^{(0)} = \sin(k+1)$, $k = 0, \ldots, 10^3 - 1$, normalized such that $\|u^{(0)}\| = 1$. We first apply the continuous-time gradient dynamical system and saddle point dynamics for finding the $\mathcal{L}_2$-gain as well as the tightest cone containing the input-output behavior via numerical integration in MATLAB with *ode15s*. Secondly, we apply the presented iterative sampling schemes. In case of the $\mathcal{L}_2$-gain, we choose Algorithm 1 in [82]. For finding the tightest cone, we apply the *Uzawa* method (cf. Proposition 3.6) with a step size of $\alpha = 0.002$. The simulation results in Figure 3.12 confirm the convergence guarantees provided in Section 3.1. Allowing for conic relations instead of the $\mathcal{L}_2$-gain decreases the radius to $r_{\min} = 7.7$.

### 3.3.2 Shortage of passivity of an oscillating system

We next consider the oscillator given by

$$\dot{x}(t) = \begin{bmatrix} -0.1 & 1 \\ -1 & 0.1 \end{bmatrix} x(t) + \begin{bmatrix} 0 \\ 1 \end{bmatrix} u(t), \quad y(t) = \begin{bmatrix} 0 & 1 \end{bmatrix} x(t) + 0.01 u(t), \qquad (3.41)$$

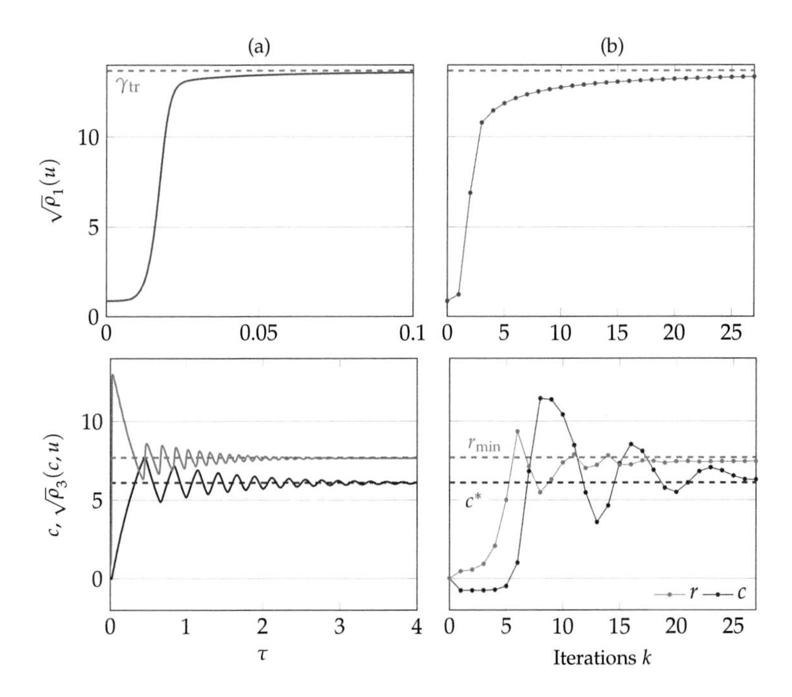


**Figure 3.12.** Illustration of the (a) continuous-time and (b) discrete-time optimization to determine the $\mathcal{L}_2$-gain (top) and the conic relation with minimal radius $(c^\star, r_{\min})$ (bottom).

in the time interval $t \in [0,10]$. We simulate the model with a sampling rate of $\Delta t = 0.01$. Its true shortage of passivity is $\beta_{\text{tr}} = 0.07$. We first apply the gradient dynamical system described in (3.16) and then apply the iterative sampling scheme including the line search algorithm in (3.22). The initial input $u^{(0)} \in \mathbb{R}^{10^3}$ is chosen to be the normed constant signal $u^{(0)} = (10\sqrt{10})^{-1} \begin{bmatrix} 1 & \cdots & 1 \end{bmatrix}^\top$. The results in Figure 3.13 reveal after seven iterations that the system is not (output strictly) passive, and we can approximate the shortage of passivity after only few iterations. However, very close to the true minimum of $\rho_2$, convergence becomes quite slow, which might be due to the fact that the generalized eigenvalues of the matrix pair $(\frac{1}{2}(G^\top + G), G^\top G)$ are spread out, which is an indicator for slow convergence of steepest descent methods (cf. (3.40) with $L_\rho \gg l_\rho$).

### 3.3.3 High-dimensional system

The third example is taken from the literature and can be found, for example, in [96] and references therein[5]. The SISO LTI model of order 84 describes the discretization of a partial differential equation (PDE) over a $7 \times 12$ grid, where the boundaries of interest lie on the opposite corners of a square. The example is listed as a benchmark example for model order reduction when the exact mathematical model is known. We simulate the trajectories with a sampling rate of $\Delta t = 5e^{-5}$ over $N = 10^4$ steps. The true $\mathcal{L}_2$-gain of the discrete-time system is $\gamma_{\text{tr}} = 10.8$ and the input feedforward passivity index is $\nu_{\text{tr}} = -0.07$. Furthermore, the measurements are subject to multiplicative noise, i.e., $\tilde{y}^{(i)} = (1 + \varepsilon^{(i)})y^{(i)}$, where $\varepsilon_k^{(i)}$ is uniformly distributed in the interval $[-\bar{\varepsilon}, \bar{\varepsilon}]$, $k = 0, \ldots, N-1$, with $\bar{\varepsilon} = 0.5$. For both, the $\mathcal{L}_2$-gain and input strict passivity, we choose the initial input to be $u_k^{(0)} = \sin(k+1) + 0.25$, $k = 0, \ldots, 10^4 - 1$, normalized such that $\|u^{(0)}\| = 1$. We apply a gradient ascent and descent, respectively, with gradient information from two noise corrupted data samples per iteration as discussed in Section 3.2.3 and we choose a fixed step size of $\alpha = 0.01$. The results in Figure 3.14 show that the presented approach converges quite fast towards a small neighborhood of the true system property despite high noise levels, which is well aligned with the discussions in Section 3.2.3.

[5] The authors of [96] made their MATLAB files available on http://verivital.com/hyst/pass-order-reduction/.

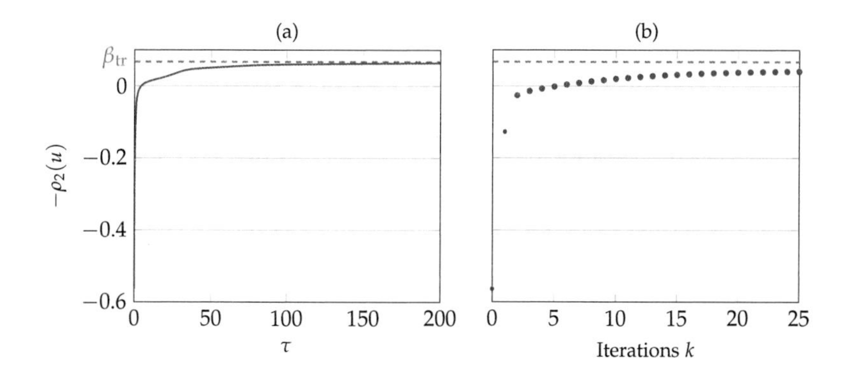


**Figure 3.13.** Illustration of the (a) continuous-time and (b) discrete-time optimization to determine the shortage of passivity $\beta$.

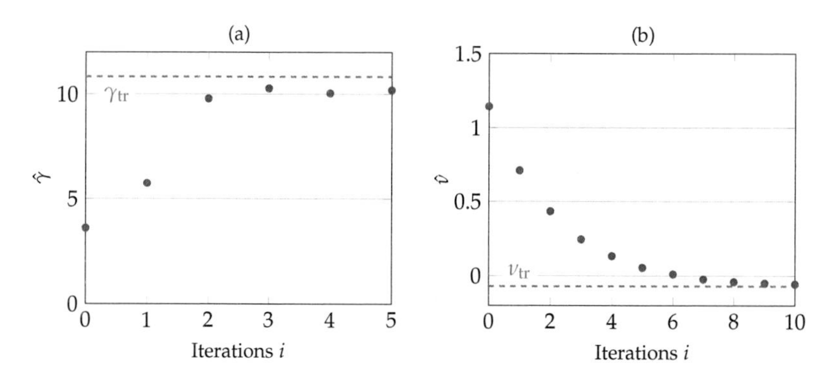


**Figure 3.14.** Iteratively determining (a) the gain and (b) passivity of the discretized PDE system with measurement noise levels at 50%.

## 3.4 Summary

In this chapter, we generalized the state-of-the-art iterative sampling scheme to find the $\mathcal{L}_2$-gain of an unknown LTI system to a broader unified framework, which not only considers determining the $\mathcal{L}_2$-gain but also input strict and output strict passivity properties as well as conic relations. First, we formulated these control-theoretic properties in terms of optimization problems, where the gradients can be obtained from input-output data samples. To find the solution to the optimization problems, we applied gradient dynamical systems and saddle point flows, whose convergence properties we investigated, and introduced tailored sampling strategies for the respective input-output system properties. Finally, we showed how this framework can be extended and generalized by introducing ideas and approaches for continuous-time systems, MIMO systems, and convergence analysis under the presence of measurement noise.

While the requirement of multiple possibly time-consuming experiments can generally be considered a drawback of the presented approaches, the iterative sampling schemes are particularly simple to apply and the implementation as well as the provided guarantees are independent of the system order. Furthermore, they have an inherent robustness against noise, which is well-studied in the literature of gradient methods. Finally, the presented methods together with the theoretical insights to the respective optimization problems come with great potential in further developing and improving its scheme. One particularly interesting direction would be to consider (slightly) nonlinear systems since [38] illustrated by means of a simulation example that the power iteration method can provide good estimates on the $\mathcal{L}_2$-gain for certain nonlinear systems. Initial ideas for a rigorous analysis of (gradient-based) sampling strategies for data-driven analysis of nonlinear systems include tailored robust gradient schemes (cf. [60]), gradient schemes for Hamiltonian systems (cf. [28]), or discrete-time extremum seeking approaches (cf. [25]).

# Chapter 4

# Offline approaches to determine input-output properties

One of the limitations of the sampling schemes to infer input-output properties from data in the previous chapter is that requiring iterative experiments can be quite restrictive and possibly time-consuming in practice. Therefore, we consider in this chapter an offline approach, which requires only one input-output trajectory to determine system properties of unknown LTI systems. The only requirement with respect to this input-output trajectory is that the input is persistently exciting of suitable order (cf. Definition 2.3). The approach here is unitary in the sense that we provide one condition for MIMO LTI systems of the form (2.1) to satisfy an IQC (cf. Section 2.3), in contrast to the iterative approaches in the previous chapter, which required tailored algorithm design and analysis for specific classes of dissipativity or LTI systems.

More specifically, the approach in this chapter is based on Willems' fundamental lemma, originally developed in the context of behavioral systems theory in [108], which provides a characterization of all trajectories of an unknown LTI system on the basis of a single input-output trajectory. This result was reformulated in [9] as follows, providing a simple equivalent characterization of Definition 2.2 from data.

**Theorem 4.1** ([108]). *Suppose $\{u_k, y_k\}_{k=0}^{N-1}$ is a trajectory of System (2.1), where $u$ is persistently exciting of order $L+n$. Then, $\{\bar{u}_k, \bar{y}_k\}_{k=0}^{L-1}$ is a trajectory of System (2.1) if and only if there exists an $\alpha \in \mathbb{R}^{N-L+1}$ such that*

$$\begin{bmatrix} H_L(u) \\ H_L(y) \end{bmatrix} \alpha = \begin{bmatrix} \bar{u} \\ \bar{y} \end{bmatrix}, \tag{4.1}$$

*where* $H_L(u)$ *and* $H_L(y)$ *denote the Hankel matrix of the input sequence* $\{u_k\}_{k=0}^{N-1}$ *and output sequence* $\{y_k\}_{k=0}^{N-1}$*, respectively, as defined in* (2.2).

This theorem basically describes that any input-output trajectory of System (2.1) can be constructed from time-shifts and linear combinations of one measured trajectory of that system. The only requirement is that the input signal of the measured trajectory is persistently exciting of sufficient order, i.e., it entails sufficient information (cf. Definition 2.3). Therefore, this result constitutes a natural basis for data-driven inference of input-output properties.

Compared to the literature on iterative sampling schemes to infer input-output properties, Willems' fundamental lemma allows to optimize over all trajectories of the system offline, while only one trajectory is available. This concept, which we will build upon in this chapter, is much in line with [57], where the idea to determine dissipativity from a single trajectory on the basis of Willems' fundamental lemma in the behavioral framework is introduced. However, their approach resulted in a nonconvex indefinite quadratic program, which is generally very hard to solve. In this chapter, on the contrary, we provide necessary and sufficient conditions for a discrete-time LTI system to satisfy an IQC, based only on a simple definiteness condition on a single data-dependent matrix.

We start in Section 4.1 by stating the main result on how to verify an $L$-IQC given one input-output trajectory of an unknown system together with a practical approach to handle measurement noise in the data. In Section 4.2, we discuss how these methods can be used to find an 'optimal' IQC to receive a more informative and tighter description of the unknown system. While Chapter 3 and most of Chapter 4 consider input-output properties of the unknown system over a finite-time horizon, we infer bounds on the respective property over the infinite-time horizon from the previous considerations in Section 4.3. Finally, we demonstrate the practicality of the introduced method on a high dimensional numerical example in Section 4.4 and conclude this chapter with a summary in Section 4.5.

This chapter is is based on and taken in parts literally from [AK4][1], [AK16][2].

[1] A. Koch, J. Berberich, J. Köhler, F. Allgöwer. "Determining optimal input-output properties: A data-driven approach." In: *Automatica* 134 (2021). p. 109906 © 2021 Elsevier Ltd.

[2] A. Romer, J. Berberich, J. Köhler, F. Allgöwer. "One-shot verification of dissipativity properties from input-output data." In: *IEEE Control Systems Letters* 3.3 (2019). pp. 709–714 © 2019 IEEE.

## 4.1 Data-driven characterization of input-output properties

In this section, we provide necessary and sufficient data-based conditions for an LTI system to satisfy an $L$-IQC. The underlying idea is to replace all possible trajectories of an LTI system in the IQC condition stated in (2.13) by a Hankel matrix containing the measured data making use of Theorem 4.1. For this purpose, we first define $\{w_k\}_{k=0}^{N-1}$ to be a stacked input-output trajectory $\{u_k, y_k\}_{k=0}^{N-1}$ as defined by

$$w_k = \begin{bmatrix} u_k \\ y_k \end{bmatrix}, \; k = 0, 1, \dots, N-1.$$

Since (2.13) only needs to hold for trajectories with $x_0 = 0$, we consider trajectories satisfying $w_0 = \dots = w_{l-1} = 0$ for some integer $l$. The restriction to this subspace can be equivalently formulated by $\tilde{V}^l w = 0$ with

$$\tilde{V}^l = \begin{bmatrix} I_{(m+p)l} & 0_{(m+p)l \times (m+p)(L-l)} \end{bmatrix} \in \mathbb{R}^{l(m+p) \times L(m+p)}.$$

Further, we define $V_L^l(w) = \left(\tilde{V}^l H_L(w)\right)^{\perp}$ to capture the condition $\tilde{V}^l w = 0$ via Finsler's lemma in the following theorem. To state necessary and sufficient conditions for an $L$-IQC from data, we let $T_L(\Psi)$ denote the block Toeplitz matrix representing the input-output map of length $L$ of a transfer function $\Psi \in \mathcal{RL}_\infty^{n_r \times (m+p)}$ with

$$T_L(\Psi) = \begin{bmatrix} g_0^\Psi & 0 & 0 & \dots & 0 \\ g_1^\Psi & g_0^\Psi & 0 & \dots & 0 \\ \vdots & & \ddots & & \vdots \\ g_{L-1}^\Psi & g_{L-2}^\Psi & \dots & & g_0^\Psi \end{bmatrix}, \tag{4.2}$$

where $\{g_k^\Psi\}_{k=0,1,2,\dots}$ is the impulse response of $\Psi$ with $g_k^\Psi \in \mathbb{R}^{n_r \times (m+p)}$.

**Theorem 4.2.** *Suppose $\{u_k, y_k\}_{k=0}^{N-1}$ is a trajectory of System (2.1).*

*(i) If $u$ is persistently exciting of order $L+n$ and*

$$V_L^{l\top}(w) H_L^\top(w) T_L^\top(\Psi)(I_L \otimes M) T_L(\Psi) H_L(w) V_L^l(w) \succeq 0 \tag{4.3}$$

*for some $l < L$, then System (2.1) satisfies the $(L-l)$-IQC defined by $P(z) = \Psi^\sim(z) M \Psi(z)$.*

(ii) *If System* (2.1) *satisfies the* $(L-l)$*-IQC defined by* $P(z) = \Psi^{\sim}(z)M\Psi(z)$*, then* (4.3) *holds for any* $l$ *with* $\underline{l} \leq l < L$*, where* $\underline{l}$ *is the lag of the system (cf. Definition 2.1).*

*Proof.*

(i) By applying Finsler's lemma [71, Lemma 2], Inequality (4.3) is equivalent to

$$\alpha^\top H_L^\top(w) T_L^\top(\Psi)(I_L \otimes M) T_L(\Psi) H_L(w)\alpha \geq 0 \tag{4.4}$$

for all $\alpha \in \mathbb{R}^{N-L+1}$ that satisfy $\tilde{V}^l H_L(w)\alpha = 0$. With Theorem 4.1, this yields

$$\bar{w}^\top T_L^\top(\Psi)(I_L \otimes M) T_L(\Psi)\bar{w} \geq 0 \tag{4.5}$$

for all trajectories $\{\bar{w}\}_{k=0}^{L-1}$ of System (2.1) with $\bar{w}_0 = \cdots = \bar{w}_{l-1} = 0$. This in turn implies

$$\bar{w}'^\top T_{L-l}^\top(\Psi)(I_{L-l} \otimes M) T_{L-l}(\Psi)\bar{w}' \geq 0 \tag{4.6}$$

for all trajectories $\{\bar{w}'\}_{k=0}^{L-l-1}$ of System (2.1) with zero initial condition.

As $\Psi$ is an LTI system with zero initial condition, the output of $\Psi$ for any given input $\{\bar{w}'\}_{k=0}^{L-l-1}$ can be cast in matrix notation as $r = T_L(\Psi)\bar{w}'$, which is hence equivalent to the convolution in (2.12). Therefore, we obtain

$$\sum_{k=0}^{L-l-1} r_k^\top M r_k \geq 0, \quad \text{with } r_k = \left(g^\Psi * \bar{w}'\right)_k, \tag{4.7}$$

which, together with the result of Proposition 2.2, results in System (2.1) satisfying the $(L-l)$-IQC as stated in Definition 2.7.

(ii) Satisfying an $(L-l)$-IQC (with $L > l$) implies that (4.7) and hence (4.6) holds for all $\{\bar{w}'\}_{k=0}^{L-l-1}$ of System (2.1) with zero initial condition. For any trajectory $\{\bar{w}_k\}_{k=0}^{L-1}$ of System (2.1), $\tilde{V}^l\bar{w} = 0$ together with $\underline{l} \leq l$ imply $x_l = 0$. Therefore, (4.6) for all $\{\bar{w}'\}_{k=0}^{L-l-1}$ of System (2.1) with zero initial condition implies that (4.5) holds for all trajectories $\{\bar{w}\}_{k=0}^{L-1}$ of System (2.1) with $\bar{w}_0 = \cdots = \bar{w}_{l-1} = 0$ whenever $\underline{l} \leq l$. Since $H_L(w)\alpha$ with $\alpha \in \mathbb{R}^{N-L+1}$ such that $\tilde{V}^l H_L(w)\alpha = 0$ is a subset of all trajectories $\{\bar{w}\}_{k=0}^{L-1}$ of System (2.1) with $\bar{w}_0 = \cdots = \bar{w}_{l-1} = 0$ (with equality if $u$ is persistently exciting of order $L+n$), Inequality (4.4) follows from (4.5). Applying Finsler's lemma similarly to Part (i) (as Finsler's lemma provides necessary and sufficient conditions), this in turn yields (4.3).

■

The above theorem provides a particularly simple approach to certify $L$-IQCs from data. Firstly, while Definition 2.6 requires that *all* possible input-output trajectories satisfy the Inequality (2.11), the condition in Theorem 4.2 is based on only *one* measured input-output trajectory of the system. Secondly, the condition to certify an $L$-IQC defined by a multiplier $P$ finally boils down to simply verifying a semidefiniteness condition on one matrix given in (4.3), which can be obtained from the given input-output trajectory. This semidefiniteness condition can be simply verified by computing, for example, the smallest eigenvalue of the resulting matrix via MATLAB functions such as *eigs*, and it is additionally also intuitive to understand: While $T_L^\top(\Psi)(I_L \otimes M)T_L(\Psi)$ represents the IQC multiplier under consideration, the Hankel matrix $H_L(w)$ spans the system behavior and $V_L^l(w)$ relaxes the conditions to trajectories with zero initial conditions. Theorem 4.2 also naturally includes necessary and sufficient conditions for $L$-dissipativity as a special case, which can also be shown in the behavioral framework [110].

There are only two requirements for verifying an $L$-IQC via Theorem 4.2. The sufficient condition in Part (i) of Theorem 4.2 requires persistence of excitation of the input of order $L+n$ to ensure that the Hankel matrix spans the *full* system behavior. Note that this persistency of excitation condition implies $N \geq (m+1)(L+n)-1$ (cf. Section 2.1), while in Chapter 3 the considered horizon $L$ of the input-output property is equal to the data length $L = N$. If the image of $H_L(w)$ does, however, not span the full system behavior, any vector in the image of $H_L(w)$ is still a trajectory of the underlying LTI system. Hence, even if the input signal of the available data pair $\{u_k, y_k\}_{k=0}^{N-1}$ is not persistently exciting, we can nevertheless verify via Part (ii) of Theorem 4.2 whether a system does *not* satisfy a specific $L$-IQC.

The other requirement in Theorem 4.2 is knowledge of an upper bound on the lag of the system $\underline{l}$ denoted by $l$. More specifically, this upper bound $l$ is required for the necessary condition in Part (ii) of Theorem 4.2. It is used in the matrix $V_L^l$, which restricts the IQC condition to trajectories with zero initial conditions. Definition 2.7 thus directly explains the requirement of $l \geq \underline{l}$ in Part (ii) of Theorem 4.2. We note that $l$ being an upper bound on the lag $\underline{l}$ is not necessary for Part (i) of Theorem 4.2, however, choosing $l$ lower than $\underline{l}$ might result in a violation of (4.3) even though the system in fact satisfies the given IQC. As an alternative to an upper bound on the lag $\underline{l}$ of the system, one can select $l$ as an upper bound on the system order $n$, since $\underline{l} \leq n$ always holds. For practical applications, $l$ can simply be chosen relatively large with the only drawback that the horizon over which an $L$-IQC is guaranteed, i.e., $L - l$, decreases.

**Remark 4.1.** Throughout the chapter, we assume $\Psi$ to be causal and stable for simplicity, as there always exists a factorization of $P$ such that this is satisfied [41, Lemma 1]. However, Theorem 4.2 can also be extended to acausal multipliers $\Psi$ as Toeplitz matrices can represent causal and acausal LTI operators. Special attention must then be given to the choice of $l$. For more details on handling acausal multipliers via Toeplitz matrices, the reader is referred to [60].

**Remark 4.2.** While Theorem 4.2 includes the filter $\Psi$ as a Toeplitz matrix $T_L(\Psi)$ multiplied to $I_L \otimes M$, one could also filter the measured trajectory $\{u_k, y_k\}_{k=0}^{N-1}$ by $\Psi$ and apply the results from [AK16] to the filtered signal $\{r_k\}_{k=0}^{N-1}$. However, one would need to account for the order of the filter in the estimate of $l$. More importantly, this would not allow to optimize over the filter $\Psi$, which will be done in Section 4.2.

Compared to iterative methods for data-driven estimation of system properties (cf. Chapter 3), the proposed method can verify a certain input-output property in only one iteration. One of the possible benefits of iterative methods is the fact that, in addition to the computation of the optimal supply rate, they also construct the corresponding worst-case trajectory, which induces the optimal supply rate. Using suitable eigenvectors of the matrix in (4.3), Theorem 4.2 can also be used to directly construct such trajectories, as the following result shows. For ease of notation, we denote the matrix in the semidefiniteness condition in (4.3) by

$$D_L^l(w) = V_L^{l\top}(w) H_L^\top(w) T_L^\top(\Psi)(I_L \otimes M) T_L(\Psi) H_L(w) V_L^l(w).$$

**Proposition 4.1.** *Suppose $\{u_k, y_k\}_{k=0}^{N-1}$ is a trajectory of System (2.1). Let $v$ be an eigenvector of $D_L^l(w)$ and define $\{\bar{w}_k\}_{k=-l+1}^{L-l-1}$ to be the trajectory of System (2.1) given by $\bar{w} = H_L(w) V_L^l(w) v$. If $v^\top D_L^l(w) v < 0$ and $l \geq \underline{l}$, then the trajectory $\{\bar{w}_k\}_{k=-l+1}^{L-1}$ satisfies $\bar{x}_0 = 0$ in any minimal realization and*

$$\sum_{k=0}^{L-l-1} r_k^\top M r_k < 0 \quad \text{with } r_k = \left(g^\Psi * \bar{w}\right)_k.$$

*Proof.* Note that from $\tilde{V}^l\bar{w} = \tilde{V}^l H_L(w)\left(\tilde{V}^l H_L(w)\right)^{\perp} v = 0$ it follows that $\bar{w}_{-l+1} = \cdots = \bar{w}_0 = 0$ and thus $\bar{x}_0 = 0$. The remainder of the proof follows the arguments in the proof of Theorem 4.2, where again Theorem 4.1 is used to identify trajectories of System (2.1) from the available input-output trajectory $\{u_k, y_k\}_{k=0}^{N-1}$ via the span of the Hankel matrix $H_L(w)$. ■

Proposition 4.1 shows that, if $D_L^l(w)$ has a negative eigenvalue and $l \geq \underline{l}$, i.e., the system does not satisfy the given ($L$-)IQC, then Theorem 4.2 can be used to construct a trajectory which violates the IQC condition. This insight can also be used to construct trajectories which attain the optimal supply rate, such as, e.g., the $\mathcal{L}_2$-gain of an unknown system. For instance, if the $\mathcal{L}_2$-gain is equal to $\gamma$, then a trajectory attaining this bound approximately can be constructed as a violating trajectory for the supply rate $\Pi_\gamma$ in (2.5) with $\gamma' = \gamma - \epsilon$ for some small $\epsilon > 0$.

### 4.1.1 Noisy measurements

Theorem 4.2 provides an equivalent data-based $L$-IQC characterization given that the measurements $\{u_k, y_k\}_{k=0}^{N-1}$ are indeed an input-output trajectory of an unknown System (2.1). In practice, however, the output of System (2.1) can in many cases not be measured exactly, but might be subject to measurement noise. Simple examples reveal that slight perturbations of the measured output $\{y_k\}_{k=0}^{N-1}$ lead to a violation of the semidefiniteness condition in (4.3), even when the unknown system does in fact satisfy the $L$-IQC of interest. Thus, for a practical scenario, the application of Theorem 4.2 needs to be refined.

In this section, we consider $L$-IQCs in the presence of measurement noise of the form

$$\tilde{y}_k = y_k + \varepsilon_k, \quad k = 0, \ldots, N-1,$$

where $\{\varepsilon_k\}_{k=0}^{N-1}$ is an instance of a stochastic process. An intuitive extension of Theorem 4.2 would be to check (4.3) with $y_k = \tilde{y}_k - \varepsilon_k$ for multiple additional noise samples $\varepsilon_k$, which can be drawn offline, in order to prove or disprove $L$-IQCs with high probability. However, as mentioned above, slightly perturbing the output in (4.3) already leads to a violation of the nonstrict matrix inequality. Therefore, we pursue a different approach, which is based on a simple, heuristic relaxation of (4.3). While this approach does not imply any guarantees on whether the unknown system satisfies specific IQCs, it works quite well in practice even for low signal-to-noise ratio (SNR) levels, as we will show in multiple numerical examples.

Although low measurement noise levels suffice to render the matrix in (4.3) indefinite, it can also be observed that the negative eigenvalues induced by the noise are by magnitudes smaller (in absolute value) than negative eigenvalues resulting from system properties, i.e., multipliers, which the system does not satisfy. With this observation in mind, an intuitive refinement of the noise-free condition is to relax (4.3) to the condition

$$V_L^{l\top}(\tilde{w})H_L^\top(\tilde{w})T_L^\top(\Psi)(I_L \otimes M)T_L(\Psi)H_L(\tilde{w})V_L^l(\tilde{w}) \succeq \delta I \tag{4.8}$$

for some $\delta < 0$, $|\delta|$ small, where $\tilde{w}_k = \begin{bmatrix} u_k^\top & \tilde{y}_k^\top \end{bmatrix}^\top$, $k = 0, \ldots, N-1$. Although this condition has no rigorous implications on (4.3), one would expect that, for small noise levels, it should serve as a good approximation of the original IQC condition.

In the following, we derive a procedure to choose $\delta$ depending on the measurements as well as samples from the noise distribution. Define

$$\begin{aligned} E_L(\tilde{w}) &= H_L^\top(\tilde{w})T_L^\top(\Psi)(I_L \otimes M)T_L(\Psi)H_L(\tilde{w}), \\ F_L(\tilde{w}, \varepsilon) &= E_L(\tilde{\tilde{w}}) - E_L(\tilde{w}), \end{aligned}$$

with $\tilde{\tilde{w}}_k = \tilde{w}_k + (0 \;\; \varepsilon_k^\top)^\top, k = 0, \ldots, N-1$. The idea is, on a high level, to approximate the perturbation that the noise causes on the matrix condition from noisy data $\{\tilde{w}_k\}_{k=0}^{N-1}$, and relax the semidefiniteness condition in (4.8) accordingly. Intuitively, $F_L(\tilde{w}, \varepsilon)$ can be viewed as the perturbation, which the noise causes onto the matrix condition in (4.3), when neglecting the kernel matrix $V_L^l(\tilde{w})$. In practice, $F_L(\tilde{w}, \varepsilon)$ is usually indefinite, and hence, loosely speaking, its negative eigenvalues cause the violation of the semidefiniteness condition in (4.3) for noisy measurements. Thus, we propose the relaxation that $E_L(\tilde{w})$ does not need to be positive semidefinite on the span of $V_L^l(\tilde{w})$, but only larger than the lowest eigenvalue of $F_L(\tilde{w}, \varepsilon)$. With (4.8), this yields

$$\delta = \lambda_{\min}\left(V_L^l(\tilde{w})^\top F_L(\tilde{w}, \varepsilon) V_L^l(\tilde{w})\right). \tag{4.9}$$

In practice, the true noise instance $\varepsilon$ is not available and $\delta$, as given in (4.9), can thus not be computed exactly. However, we can sample an arbitrary noise instance from the noise distribution offline and approximate $\delta$ in (4.9) with this noise instance. To reduce the influence of a specific noise instance, we can sample $K$ arbitrary noise instances $\{\varepsilon_k^{(i)}\}_{k=0}^{N-1}$, $i = 1, \ldots, K$ from the assumed noise distribution offline and compute $\delta$ by

$$\delta = \frac{1}{K} \sum_{i=1}^{K} \lambda_{\min} \left( V_L^{l\top}(\tilde{w}) F_L(\tilde{w}, \varepsilon^{(i)}) V_L^l(\tilde{w}) \right). \tag{4.10}$$

For infinitely many noise instances, we would thus retrieve the expected value of $\delta$ for the given distribution. While knowledge of the exact noise distribution is usually not available, using an approximation of this distribution still provides an indication on how to choose $\delta$ for a reasonable relaxation.

To summarize, we extend the developed $L$-IQC condition to a measurement noise setting by introducing the relaxation in (4.8), where a reasonable value of $\delta$ can be obtained via (4.10). The procedure is summarized in the following algorithm.

**Algorithm 1: IQCs from noisy measurements**

1. Measure data $\{u_k, \tilde{y}_k\}_{k=0}^{N-1}$.
2. Draw $K$ noise samples $\{\varepsilon_k^{(i)}\}_{k=0}^{N-1}$, $i = 1, \ldots, K$ offline from the noise distribution.
3. Compute $\delta$ using (4.10).
4. Use (4.8) for checking the $L$-IQC.

If no prior knowledge on the noise distribution is available, one can also attempt to learn a noise distribution from data. Ideas in this direction can be found, e.g., in [6].

**Remark 4.3.** Although we have only included measurement noise into the analysis above, we could also try to reduce the effect of measurement noise in the data. If we have access to additional data tuples, we can reduce the variance of the measurement noise by averaging over multiple trajectories (or taking any convex combination of trajectories). This is again a trajectory of the unknown system and, provided that the averaged input is persistently exciting, can be used to verify $L$-IQCs of the underlying system via the presented approach. This can lead to better results since averaging of output trajectories can reduce the variance in the noise significantly. If there are no additional trajectories available, one could also split a single measured data trajectory into multiple parts of shorter lengths to build the average of these trajectories. This leads to a trade-off between reducing the variance of the measurement noise and the considered time horizon $L$ given a fixed amount of input-output data. Averaging or any other methods to de-noise the data can easily be combined with the above algorithm.

While the above approach does not provide guarantees on the resulting IQC, it provides very promising results in practice, as will be illustrated in the following example.

**Example 4.1.** We demonstrate the applicability of Algorithm 1 with a numerical example. To this end, we choose a random $2 \times 2$ MIMO system with system order $n = 3$ via the MATLAB function *drss(3,2,2)* with the seed *rng(2)*. We are interested in the following IQC similar to [27, p. 3147]:

$$\Psi(z) = \begin{bmatrix} B_1(z) \otimes I_m & 0 \\ 0 & B_2(z) \otimes I_p \end{bmatrix}, \quad M = \begin{bmatrix} \gamma^2 I_m & 0 \\ 0 & -I_{3p} \end{bmatrix},$$

where $B_1(z) = 1$ and $B_2(z) = \begin{bmatrix} 1 & \frac{1}{z-0.5} & \frac{1}{(z-0.5)^2} \end{bmatrix}^\top$.

We consider trajectories of the length $N = 1000$ with the output being subject to uniform multiplicative noise of the form $\tilde{y}_k = (1 + \varepsilon_k) y_k$ with $\varepsilon_k \in [-\bar{\varepsilon}, \bar{\varepsilon}]$ where $\bar{\varepsilon} > 0$ represents the SNR. For each noise level $\bar{\varepsilon}$ we draw 10 different noise sequences affecting the measured output.

For each of the noisy input-output trajectories, we first apply Algorithm 1 together with bisection to find the smallest $\gamma$ such that the unknown system still satisfies the $L$-IQC. For this, we choose $L = 100$, $K = 3$ and assume an upper bound on the lag of $l = 5$. For comparison, we consider also the MATLAB function *ssregest* from the System Identification Toolbox [47] with consecutive systems analysis to find the smallest $\gamma$. In particular, we apply the system identification function *ssregest*, which estimates a state-space model by reduction of a regularized ARX model, and we assume knowledge of the exact model order $n = 3$.

For both approaches, we calculate the normalized error $e_\gamma = \frac{|\gamma - \hat{\gamma}|}{\gamma}$ for each resulting estimate $\hat{\gamma}$. At each noise level $\bar{\varepsilon}$, we calculate the mean of the normalized error $e_\gamma$ from the 10 estimates for each method. The result can be seen in Table 4.1.

While, very generally, the results from both approaches deteriorate with larger noise levels $\bar{\varepsilon}$, this small example supports the claim that the presented approach approximates the respective system property even for large noise levels. Further,

**Table 4.1.** Mean of the normalized error $e_\gamma$ at different noise levels resulting from Algorithm 1 and a standard system identification method with subsequent model-based analysis, respectively.

| $\bar{\varepsilon}$ | 0.1 | 0.15 | 0.2 | 0.25 | 0.3 |
|---|---|---|---|---|---|
| Algorithm 1: mean $e_\gamma$ (%) | 3.3 | 4.5 | 5.4 | 5.3 | 6.4 |
| *ssregest*: mean $e_\gamma$ (%) | 2.5 | 2.5 | 4.6 | 6.4 | 8.0 |

the error range is similar to the error from a standard system identification method with known model order $n$ and consecutive systems analysis. The advantages of the presented method (being nonparametric and not requiring exact knowledge of the system order) become more apparent for high dimensional systems as demonstrated and discussed in Section 4.4.

**Remark 4.4.** It is an interesting and largely open issue for future research to extend the results of this chapter to provide rigorous guarantees in the case of measurement noise. However, guarantees from noisy data of finite length are very generally and also in system identification approaches still part of current research, typically requiring strong assumptions. Two very interesting recent results in this direction are, e.g., [91] for nonasymptotic system identification results from input-state measurements with zero initial condition and Gaussian i.i.d. noise on the state measurements, or [74] for guarantees from input-output measurements of finite length given Gaussian i.i.d. noise and a Gaussian i.i.d. input, known system order $n$, and zero initial condition assumption.

Alternatively, in the sense of optimal experiment design, the proposed approach together with Proposition 4.1 can also be used to calculate an input trajectory close to the 'optimal' one in the sense that this input trajectory corresponds to the worst-case trajectory with respect to a certain system property. Many iterative schemes operate with the aim to find exactly this input trajectory (cf. Chapter 3, [63, 106]). Applying the results of this section to initialize such iterative schemes with the estimated 'optimal' input trajectory has the potential to immensely speed up their convergence. Such iterative schemes commonly provide qualitative error bounds under the presence of noise (cf. Sec. 3.2.3). Combining the two approaches thus provides a systematic approach for obtaining guarantees in the noisy case with increased data efficiency.

## 4.2 Data-driven inference of optimal input-output properties

In the previous section, we introduced an approach to verify whether an LTI system satisfies an IQC for a given multiplier $P(z) = \Psi^{\sim}(z)M\Psi(z)$ (cf. Theorem 4.2). Usually, however, we are interested in finding some 'optimal' IQC, which provides the tightest description of the unknown system within a class of IQCs. For instance, one might want to estimate the $\mathcal{L}_2$-gain of the system, i.e., the minimal $\gamma$ such that the system is dissipative w.r.t. $P = \Pi_\gamma$ from (2.5). This can be done via a standard bisection method, or we can state it as a simple SDP reading

$$\min_{\gamma^2} \gamma^2 \quad \text{s.t. } V_L^{l^\top}(w)H_L^\top(w)(I_L \otimes \Pi_\gamma)H_L(w)V_L^l(w) \succeq 0.$$

Other simple but important system properties are input strict and output strict passivity. The excess or shortage of the respective passivity property, i.e., the maximal $\nu$ and the minimal $\beta$ such that the system is dissipative w.r.t. $P = \Pi_\nu$ and $P = \Pi_\beta$ from (2.5), can again be found via a simple bisection algorithm or an SDP, since $P$ is linear in $\nu$ and $\beta$, respectively.

The optimal IQC within a specific class of parameterized IQCs that a system satisfies in case of the $\mathcal{L}_2$-gain or passivity parameters is quite intuitive. Similarly, we consider a more general optimal IQC within a specified parameterized class of IQCs to be an IQC for which the semidefiniteness condition is tight. For an optimal IQC, there hence exists a nontrivial input-output tuple $\{u_k, y_k\}_{k=0}^{L-1}$ for which (2.11) holds with equality. In this sense, the optimal IQC represents a tight description within the given class of IQCs.

A very important class of IQCs that has been extensively studied in literature are positive-negative (PN) multipliers for which there exists a factorization such that

$$M_\gamma = \begin{bmatrix} \gamma^2 I_{n_{r_1}} & 0 \\ 0 & -I_{n_{r_2}} \end{bmatrix}, \quad \Psi = \begin{bmatrix} \Psi_{11} & \Psi_{12} \\ \Psi_{21} & \Psi_{22} \end{bmatrix}, \tag{4.11}$$

with $\Psi_{11} \in \mathcal{RH}_\infty^{n_{r_1} \times m}$, $\Psi_{12} \in \mathcal{RH}_\infty^{n_{r_1} \times p}$, $\Psi_{21} \in \mathcal{RH}_\infty^{n_{r_2} \times m}$, and $\Psi_{22} \in \mathcal{RH}_\infty^{n_{r_2} \times p}$. Let us further only consider filters $\Psi$ for which $\Psi_{12} = 0$ (cf. triangular factorization for positive-negative multipliers as discussed in [18]), $\Psi_{11}(z)$ fixed, and $\Psi_{21}(z)$, $\Psi_{22}(z)$

linearly parameterized, i.e.,

$$\Psi(z) = \begin{bmatrix} \Psi_{11}(z) & 0 \\ \Psi_{21}(z) & \Psi_{22}(z) \end{bmatrix}, \tag{4.12}$$

with

$$\Psi_{21}(z) = \sum_{k=0}^{b} c_k^{(21)} B_k^{(21)}(z), \quad \Psi_{22}(z) = \sum_{k=0}^{b} c_k^{(22)} B_k^{(22)}(z). \tag{4.13}$$

Here, $B_k^{(21)}(z)$, $B_k^{(22)}(z)$, $k = 0, \ldots, b$ are fixed basis functions and $c_k^{(21)} \in \mathbb{R}^{n_{r_2} \times m}$, $c_k^{(22)} \in \mathbb{R}^{n_{r_2} \times p}$ are free parameters. Popular basis functions include, for example, $(1, (z+\lambda)^{-1}, (z+\lambda)^{-2}, \ldots, (z+\lambda)^{-b})$, with $|\lambda| < 1$ fixed. This specific choice of basis functions has the beneficial property that any $\Psi \in \mathcal{RH}_\infty$ can be approximated arbitrarily closely in the $\mathcal{H}_\infty$-norm by a suitable linear combination if $b$ is chosen sufficiently large (cf. Remark 5 in [100] and the references therein). The corresponding Toeplitz matrices of $\Psi_{21}$ and $\Psi_{22}$ can be computed as

$$T_L(\Psi_{21}) = \sum_{k=0}^{b} T_L(c_k^{(21)} B_k^{(21)}), \quad T_L(\Psi_{22}) = \sum_{k=0}^{b} T_L(c_k^{(22)} B_k^{(22)}).$$

Note that $T_L(c_k^{(21)} B_k^{(21)})$ and $T_L(c_k^{(22)} B_k^{(22)})$ are block Toeplitz matrices, possibly non-square. In the following, we present a semidefinite program (SDP) which computes the optimal IQC within the aforementioned class of IQCs that an a priori unknown LTI system satisfies from only one input-output trajectory.

To improve readability of the main results, we rearrange vectors and matrices and denote

$$\tilde{\tilde{V}}^l = \begin{bmatrix} I_{ml} & 0_{ml \times m(L-l)} & 0_{ml \times pl} & 0_{ml \times p(L-l)} \\ 0_{pl \times ml} & 0_{pl \times m(L-l)} & I_{pl} & 0_{pl \times p(L-l)} \end{bmatrix},$$

$$H_u^l = H_L(u) \left( \tilde{\tilde{V}}^l \begin{bmatrix} H_L(u) \\ H_L(y) \end{bmatrix} \right)^\perp, \quad H_y^l = H_L(y) \left( \tilde{\tilde{V}}^l \begin{bmatrix} H_L(u) \\ H_L(y) \end{bmatrix} \right)^\perp,$$

and $T_{11} = T_L(\Psi_{11})$, $T_{21}(c) = T_L(\Psi_{21}(c^{21}))$, $T_{22}(c) = T_L(\Psi_{22}(c^{22}))$ to emphasize the parameter dependence of $\Psi_{21}$ and $\Psi_{22}$.

We are now interested in finding the minimal $\gamma^2$ over all parameterized multipliers given by (4.11)-(4.13) such that the $L$-IQC holds for the unknown LTI system on the basis of only one input-output trajectory.

**Theorem 4.3.** *Suppose $\{u_k, y_k\}_{k=0}^{N-1}$ is a trajectory of System (2.1), $u$ is persistently exciting of order $L+n$, and $\underline{l} \leq l < L$. The smallest $\gamma^2$ such that System (2.1) satisfies the $(L-l)$-IQC with a multiplier defined by (4.11)-(4.13) can be computed by*

$$\min_{\gamma^2, c} \gamma^2 \quad s.t. \quad \begin{bmatrix} I & (T_{21}(c)H_u^l + T_{22}(c)H_y^l)^\top \\ T_{21}(c)H_u^l + T_{22}(c)H_y^l & \gamma^2 H_u^{l\top} T_{11}^\top T_{11} H_u^l \end{bmatrix} \succeq 0. \tag{4.14}$$

*Proof.* From Theorem 4.2 we know that System (2.1) satisfies an $(L-l)$-IQC defined by (4.11)-(4.13) if

$$\begin{bmatrix} H_u^l \\ H_y^l \end{bmatrix}^\top \begin{bmatrix} T_{11} & 0 \\ T_{21}(c) & T_{22}(c) \end{bmatrix}^\top M_\gamma \otimes I_L \begin{bmatrix} T_{11} & 0 \\ T_{21}(c) & T_{22}(c) \end{bmatrix} \begin{bmatrix} H_u^l \\ H_y^l \end{bmatrix} \succeq 0.$$

This yields the equivalent semidefiniteness condition

$$\gamma^2 H_u^{l\top} T_{11}^\top T_{11} H_u^l - \left[T_{21}(c)H_u^l + T_{22}(c)H_y^l\right]^\top \left[T_{21}(c)H_u^l + T_{22}(c)H_y^l\right] \succeq 0.$$

Using the Schur complement, we can rewrite this problem as the SDP given in (4.14). ∎

The resulting optimization problem in (4.14) is an SDP with $2(b+1)+1$ decision variables that can be solved efficiently using standard solvers. Note that the size of the matrix condition in (4.14) grows linearly with the length of the input-output trajectory.

One very important system property that falls into the class of IQCs defined by (4.11)-(4.13) are conic relations ($P = \Pi_c$ in (2.5)), as introduced in more detail in Section 3.1.3. In view of the factorization in (4.11), we are interested in the minimal $\gamma \geq 0$ for some matrix $C \in \mathbb{R}^{p \times m}$ such that System (2.1) satisfies an IQC (or dissipativity property) characterized by $P = \Psi_c^\top M_\gamma \Psi_c$, where

$$\Psi_c = \begin{bmatrix} I & 0 \\ -C & I \end{bmatrix}. \tag{4.15}$$

While in Section 3.1.3 convergence towards the tightest cone (minimal $\gamma$) could be obtained via iterative experiments, Theorem 4.3 provides means to compute the tightest cone describing the a priori unknown LTI system from only one input-output trajectory via a simple SDP.

**Corollary 4.1.** *Suppose $\{u_k, y_k\}_{k=0}^{N-1}$ is a trajectory of System (2.1), $u$ is persistently exciting of order $L+n$, and $\underline{l} \leq l < L$. The smallest $\gamma^2$ such that System (2.1) is confined to the cone described by $(C, \gamma)$ over the horizon $L-l$ can be computed by*

$$\min_{\gamma^2, C} \gamma^2 \quad \text{s.t.} \quad \begin{bmatrix} I & (-C_L H_u^l + H_y^l)^\top \\ -C_L H_u^l + H_y^l & \gamma^2 H_u^{l\top} H_u^l \end{bmatrix} \succeq 0$$

*with $C_L = I_L \otimes C$.*

Very generally, while the $\mathcal{L}_2$-gain might be quite large for some systems, a conic description can decrease $\gamma$ significantly, and the use of dynamic filters provides an even smaller $\gamma$.

**Example 4.2.** We choose three random $2 \times 2$ MIMO systems with system order $n = 3$ via the MATLAB function *drss(3,2,2)* with seed *rng(i)*, $i = 2, 3, 4$. We choose $l = 3$, $L = 200$ and $N = 500$, and we determine the tightest $L$-IQC considering the $\mathcal{L}_2$-gain and multipliers as given in (4.11)-(4.13) with basis functions $\left(1, (z+\lambda)^{-1}, \ldots, (z+\lambda)^{-b}\right)$, $\lambda = 0.8$, $b = 0, 1, 2, 3$, for $\Psi_{21}$, and $\Psi_{11}(z) = \Psi_{22}(z) = I_2$. Note that $b = 0$ represents conicity (cf. Corollary 4.1). The results are illustrated in Figure 4.1 and show that allowing for a conic description already significantly improves the radius $\gamma$ compared to computing the $\mathcal{L}_2$-gain and hence decreases the conservatism in the respective feedback theorem [113]. As expected, we can decrease the radius $\gamma$ even more by utilizing dynamic filters.

**Remark 4.5.** Another special case that falls in the above category of IQCs are dynamic cones, or also called 'nonlinearity measures', as explained for example in [88]. To arrive at a dynamic cone description, we choose $\Psi_{11}(z) = I_m$ and $\Psi_{22}(z) = I_p$, which yields

$$P(z) = \begin{bmatrix} \gamma^2 I - \Psi_{21}^\top(z)\Psi_{21}(z) & \Psi_{21}(z) \\ \Psi_{21}(z) & -I \end{bmatrix}.$$

The smallest radius $\gamma$ over the set of possible LTI approximations is hence a measure on how nonlinear the system at hand is. If the underlying system is linear, the same

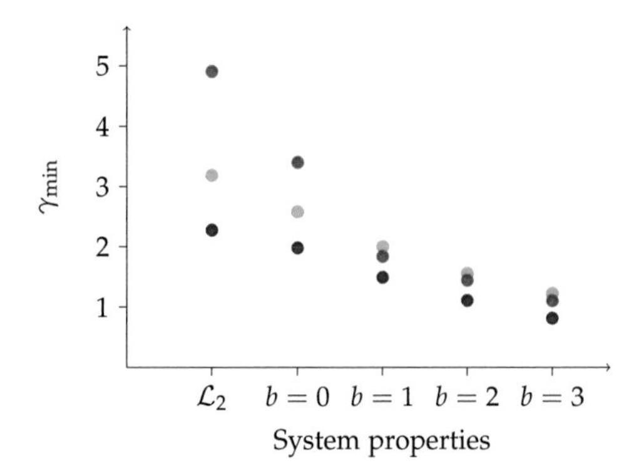


**Figure 4.1.** Data-driven inference of the $\mathcal{L}_2$-gain, tightest cone ($b = 0$), and best dynamic parameterization ($b = 1, 2, 3$) of three random systems indicated by the colors red, green, and blue, respectively.

IQC measures the distance with respect to another linear system. Such a measure can be useful in model reduction of high dimensional systems, for example, and corresponding robust controller design. Hence, for a given reduced-order model $\Psi_{21}(z)$, we can compute the approximation error $\gamma$. Furthermore, by parameterizing $\Psi_{21}(z)$ through basis functions as described above, we identify the closest LTI model within a given class. Hence, Theorem 4.3 enables what we call here lower-order model approximation (contributing towards 'data-driven model reduction' [85]). The limitation yet is that we have to parameterize the reduced model linearly to receive convergence guarantees, but with the caveat that we at the same time receive a measure $\gamma$ describing the worst-case deviation between full-order and reduced-order model.

**Example 4.3.** To underline Remark 4.5, we consider the following $7^{\text{th}}$ order system

$$\mathcal{G}_{\text{ex}}(z) = \begin{bmatrix} \frac{2}{z+0.51} & \frac{1}{z+0.19} + \frac{1}{z+0.21} \\ \frac{1}{z+0.55} + \frac{2}{z+0.2} & \frac{2}{z+0.52} + \frac{3}{z+0.5} \end{bmatrix},$$

which has an $\mathcal{L}_2$-gain of 11.9. We parameterize $\Psi_{21}(z)$ with the basis functions $(1, (z+0.5)^{-1}, (z+0.2)^{-1})$ and choose $\Psi_{11}(z) = \Psi_{22}(z) = I_2$. Choosing $l = 10$, $L = 110$ and $N = 300$, we retrieve the optimal parameterization (i.e., the minimal $\gamma$)

for

$$\Psi_{12}(z) = \begin{bmatrix} 2.1 & 0.0 \\ 1.3 & 5.2 \end{bmatrix} \frac{1}{1+0.5z} + \begin{bmatrix} -0.1 & 2.0 \\ 1.7 & 0.2 \end{bmatrix} \frac{1}{1+0.2z}$$

which closely approximates $\mathcal{G}_{\mathrm{ex}}$ with lower model order and a guaranteed approximation error of $\gamma = 0.05$.

In this section, we have introduced an SDP for finding the optimal IQC for PN multipliers as parameterized in (4.11)-(4.13), which represents a quite general and important class of IQCs. Another parameterization that is equally applicable is, e.g.,

$$M = \begin{bmatrix} 0 & I \\ I & -\gamma I \end{bmatrix}, \quad \Psi(z) = \begin{bmatrix} \Psi_{11}(z) & \Psi_{12}(z) \\ 0 & \Psi_{22}(z) \end{bmatrix},$$

(cf. [101, Eq. (30b)]), with $\Psi_{22}(z)$ a priori fixed and $\Psi_{11}(z)$, $\Psi_{12}(z)$ linearly parameterized.

## 4.3 Guarantees for the infinite horizon

Many data-based methods [57, 62, 72, 82] including the results in Chapter 3, Section 4.1, and Section 4.2 consider control-theoretic system properties holding over a finite-time horizon (cf. $L$-dissipativity and $L$-IQCs in Definition 2.5 and Definition 2.7, respectively). In this section, we therefore aim to derive an upper bound on the difference between considering system properties over the infinite and finite horizon. With such results, we aim to guarantee a (possibly conservative) system property over the infinite horizon, and investigate how the results considering the finite horizon in Definition 2.7 approach the results over the infinite horizon in Definition 2.6. To this end, we first consider IQCs satisfied by SISO systems and then explain what these results imply regarding specific system properties. Finally, we will end this section with a short discussion on MIMO systems.

We start with the following lemma that establishes a transformation that will later be used in the main result of this section.

**Lemma 4.1.** *Let $\mathcal{G}$ and $\Psi$ be causal and stable discrete-time LTI systems. If there exists a stable and causal left inverse $\Psi_{11}^{-1}$ such that $\|\Psi_{12}\mathcal{G}\Psi_{11}^{-1}\|_\infty < 1$, then $\widetilde{\mathcal{G}} : r^{[1]} \mapsto r^{[2]}$ as depicted in Figure 4.2 is causal and stable.*

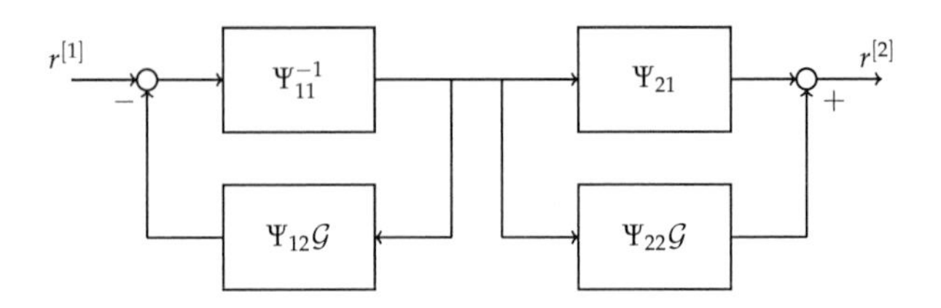


**Figure 4.2.** Transfer function $\widetilde{\mathcal{G}} : r^{[1]} \mapsto r^{[2]}$, with $r_k^{[1]} \in \mathbb{R}^{n_{r_1}}$, $r_k^{[2]} \in \mathbb{R}^{n_{r_2}}$, $k = 0, 1, \ldots$.

*Proof.* By assumption, $\Psi_{21} + \Psi_{22}\mathcal{G}$ is causal and stable. With the small-gain theorem, we know that $(\Psi_{11} + \Psi_{12}\mathcal{G})^{-1}$ is causal and stable if $\|\Psi_{12}\mathcal{G}\Psi_{11}^{-1}\|_\infty < 1$. Note that the small-gain argument also implies well-posedness of the interconnection in Figure 4.2. As a series of causal and stable LTI systems, $\widetilde{\mathcal{G}} : r^{[1]} \mapsto r^{[2]}$ as depicted in Figure 4.2 is causal and stable. ∎

First, note that assuming the existence of a stable left-inverse $\Psi_{11}^{-1}$ is not very restrictive. $\Psi_{11}$ is oftentimes a tall matrix parameterized, for example, by a certain choice of basis functions. As these basis functions usually have a constant component, there usually exists a stable left inverse $\Psi_{11}^{-1}$. Furthermore, for PN multipliers $P$, there also exists a triangular factorization where $\Psi_{11}$, $\Psi_{11}^{-1}$, $\Psi_{22}$ are stable rational transfer functions and $\Psi_{12} = 0$ [18]. For any such triangular factorization, the condition $\|\Psi_{12}\mathcal{G}\Psi_{11}^{-1}\|_\infty < 1$ is also trivially satisfied. Since the condition $\|\Psi_{12}\mathcal{G}\Psi_{11}^{-1}\|_\infty < 1$ requires the verification of a certain system property of a transformed LTI system, Theorem 4.2 can alternatively be used to check the condition from input-output data over a finite-time horizon.

Lemma 4.1 leads us to the main result of this section for SISO LTI systems, which is presented in the following.

### 4.3.1 Infinite horizon properties of SISO systems

For SISO LTI systems, where $m = p = n_{r_1} = n_{r_2} = 1$, we can bound the remainder term from an $L$-IQC to an IQC over the infinite horizon using [13, 98] via the following result.

**Theorem 4.4.** *Let $\mathcal{G}$ be a stable, discrete-time SISO LTI system which satisfies the L-IQC given by $P(z) = \Psi^{\sim}(z)M\Psi(z)$ with*

$$M = \begin{bmatrix} \gamma^2 & 0 \\ 0 & -1 \end{bmatrix},$$

*and let $\Psi$ admit a stable and causal inverse of $\Psi_{11}$ such that $\|\Psi_{12}\mathcal{G}\Psi_{11}^{-1}\|_\infty < 1$. Then, $\mathcal{G}$ satisfies the (infinite horizon) IQC given by $P(z) = \Psi^{\sim}(z)M_{\text{inf}}\Psi(z)$ with*

$$M_{\text{inf}} = \begin{bmatrix} (\gamma+\epsilon)^2 & 0 \\ 0 & -1 \end{bmatrix}$$

*and $\epsilon = \mathcal{O}\left(\frac{1}{L^2}\right)$.*

*Proof.* We introduce

$$\begin{bmatrix} r^{[1]} \\ r^{[2]} \end{bmatrix} = \begin{bmatrix} \Psi_{11} & \Psi_{12} \\ \Psi_{21} & \Psi_{22} \end{bmatrix} \begin{bmatrix} u \\ y \end{bmatrix}.$$

With $u = (\Psi_{11} + \Psi_{12}\mathcal{G})^{-1}r^{[1]}$ and $r^{[2]} = (\Psi_{21} + \Psi_{22}\mathcal{G})u$, $\widetilde{\mathcal{G}} : r^{[1]} \mapsto r^{[2]}$ can be depicted as given in Figure 4.2. Lemma 4.1 shows that the transformed system $\widetilde{\mathcal{G}}$ is causal and stable, which reduces the remainder of this proof to finding the difference of the $\mathcal{L}_2$-gain of $\widetilde{\mathcal{G}}$ over the horizon $L$ and the $\mathcal{L}_2$-gain of $\widetilde{\mathcal{G}}$ over the infinite-time horizon. With $\widetilde{\mathcal{G}}$ being causal and stable, we can apply [13, Theorem 4.1] to prove that the $\mathcal{L}_2$-gain over the horizon $L$ of $\widetilde{\mathcal{G}}$ approaches the $\mathcal{L}_2$-gain over the infinite horizon for increasing $L$ with $\epsilon = \mathcal{O}\left(\frac{1}{L^2}\right)$. ■

The general idea of Theorem 4.4 is hence to translate the question of finding the difference between an IQC over the finite and infinite horizon satisfied by a stable SISO LTI system $\mathcal{G}$ to the question of finding the difference between the $\mathcal{L}_2$-gain over the finite and infinite horizon of a transformed (but stable and causal) system $\widetilde{\mathcal{G}}$. For finding the difference between the $\mathcal{L}_2$-gain over the finite and infinite horizon of a stable and causal system, we can then directly apply the results from [13]. A specific expression for $\epsilon$ and a lower bound on $L$ to retrieve a specific $\epsilon$ can be found in [98] for all $L \geq 3$. Theorem 4.4 implies that the difference between considering a finite versus an infinite horizon vanishes at least quadratically with increasing horizon $L$.

The results above apply, for example, to the most common dissipativity properties. Choose, for example, $\Psi_c$ in (4.15) to retrieve general conic relations. This yields for the radius of the conic relation $\gamma_{\text{infinite}} \leq \gamma_{\text{finite}} + \epsilon$. Furthermore, since $\epsilon$ depends on the norm of $\widetilde{\mathcal{G}}$ [98], finding the minimal cone containing the input-output operator typically reduces also the remainder term $\epsilon$. Other examples include output strict passivity where the $\mathcal{L}_2$-gain estimation over the infinite-time horizon of the transformed system via

$$\Psi_\beta = \begin{bmatrix} 1 & 0 \\ (\gamma + \epsilon) & -\frac{1}{2(\gamma+\epsilon)} \end{bmatrix}$$

yields the output strict passivity property over the infinite-time horizon. Note that $\Psi_c$ as well as $\Psi_\beta$ immediately satisfy the conditions of Theorem 4.4, i.e., admit a stable and causal inverse of $\Psi_{11}$ and $\|\Psi_{12}\mathcal{G}\Psi_{11}^{-1}\|_\infty = 0$.

**Remark 4.6.** Theorem 4.4 combined with the results from [98] suggests roughly that smaller $\|\widetilde{\mathcal{G}}\|_\infty$ imply smaller $\epsilon$ for a given data length $L$. To decrease $\epsilon$ or in order to use shorter input-output trajectories, a very interesting filter $\Psi$ is

$$\Psi = \begin{bmatrix} I & 0 \\ 0 & \mathcal{S}^{\rho-} \end{bmatrix},$$

where $\mathcal{S}^{\rho-}$ is defined by $w^\rho = \mathcal{S}^{\rho-}w$ as $w_k^\rho = \rho^k w_k$, $\rho < 1$. For more information, the reader is referred to [42], where this filter has been used to show exponential stability. Applying these exponential filters to input-output trajectories can decrease $\epsilon$ and can extend the results of this section to unstable systems, where the numerical calculations otherwise become difficult. Special attention, however, must be paid to how this is influenced by measurement noise.

The restricting condition for the application of Theorem 4.4 to general IQCs is that we assume $n_{r_1} = n_{r_2} = 1$ although many IQC descriptions include tall filters $\Psi$. Generally speaking, $n_{r_1}, n_{r_2} > 1$ brings us to the problematic MIMO case as discussed in the next subsection.

### 4.3.2 Infinite horizon properties of MIMO systems

Since the result from [98] and hence the results above strongly rely on the Toeplitz structure of an input-output description of a discrete-time SISO LTI system and the corresponding results on the norms of Toeplitz matrices [13], they are not transferable to MIMO systems. An obvious, but quite conservative, approach is to bound the $\mathcal{L}_2$-gain over the infinite-time horizon from each of the inputs to each of the outputs and take the matrix norm as an upper bound for the $\mathcal{L}_2$-gain over the infinite-time horizon of the MIMO system. In most cases, however, this yields a quite conservative bound for the $\mathcal{L}_2$-gain of the MIMO system. Very generally, we will retrieve the exact $\mathcal{L}_2$-gain with Theorem 4.2 in the limit $L \to \infty$ also for MIMO systems, see, e.g., [81]. However, we are more particularly interested how fast we approach the true $\mathcal{L}_2$-gain (and, via transformations as introduced before, other system properties). For general MIMO systems, this remains an open question. We have a solution for the special case of quadratic MIMO finite impulse response (FIR) models, which goes back to results on norms of finite sections of perturbed Toeplitz band matrices [81]. In [81], the authors analyze how singular values of the finite section Toeplitz matrices approximate the singular values of infinite Toeplitz matrices.

**Proposition 4.2.** *Given a square FIR MIMO system $\mathcal{G}_{\text{FIR}}$,*

$$\mathcal{G}_{\text{FIR}}(z) = \sum_{k=0}^{h} g_k z^{-k}, \quad g_k \in \mathbb{R}^{m\times m},\ h \in \mathbb{N},$$

*with* $\det \mathcal{G}(z) \neq 0$ *for all $\{z \in \mathbb{C} : |z| = 1\}$. Then, for all $L \geq 20h$, L-dissipativity with*

$$P = \begin{bmatrix} \gamma^2 I_m & 0 \\ 0 & -I_m \end{bmatrix}$$

*implies dissipativity over the infinite horizon for*

$$P = \begin{bmatrix} (\gamma+\epsilon)^2 I_m & 0 \\ 0 & -I_m \end{bmatrix}$$

*with $\epsilon \leq \frac{20h}{L} \|\mathcal{G}_{\text{FIR}}\|_\infty$.*

*Proof.* This result is based on Lemma 4.3 in [81]. ∎

A more detailed explanation can be found in [AK4, Appendix A].

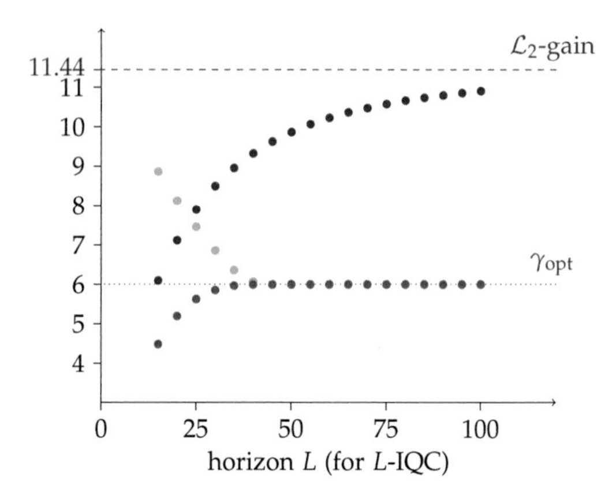


**Figure 4.3.** Determining the $\mathcal{L}_2$-gain and the tightest cone containing the input-output behavior of a $2 \times 2$ MIMO system over different horizons $L$. Here, • denotes the $\mathcal{L}_2$-gain estimate, • denotes the $\gamma$ estimate resulting from the corresponding $C$ estimate, and • denotes the true $\gamma$ with respective to the same $C$ estimate.

Together with Lemma 4.1 from before, this result has the potential to be applied also to other system properties. However, the transformed system needs to be again an FIR system, which holds for dissipativity properties but is generally not true in the case of IQCs. While the above result is quite restrictive, it exemplarily shows that even in the MIMO case some qualitative results can be obtained.

**Example 4.4.** We determine the $\mathcal{L}_2$-gain and the smallest cone containing the input-output behavior of a randomly generated $2 \times 2$ MIMO system with system order $n = 5$ (i.e., MATLAB function *drss* with *rng(0)*) over different horizons $L$. The results in Figure 4.3 illustrate different phenomena. Even though the theoretical results in Theorem 4.4 hold only for SISO systems, it seems that a similar behavior can be observed also for MIMO systems. Furthermore, by allowing for a transformation with a center matrix $C$, we can greatly reduce the conservatism of the $\mathcal{L}_2$-gain. Moreover, as suggested in the previous analysis, due to the smaller $\mathcal{L}_2$-gain of the transformed system, the radius $\gamma$ converges with increasing $L$ significantly faster towards the corresponding infinite horizon value $\gamma_{\text{opt}}$ then the $\mathcal{L}_2$-gain. During the first steps, where the optimal $C$ parameter over the finite horizon deviates from the optimal $C$ parameter over the infinite horizon, the minimal radius for the nonoptimal $C$ is larger than the minimal radius for the optimal $C$.

## 4.4 High dimensional numerical example

Throughout the last sections, we provided small examples to illustrate the presented results and effects of different parameters. In the following, we focus on a rather high dimensional system. We consider a model of a building (the Los Angeles University Hospital), which has been listed as a benchmark model reduction problem, e.g., in [19, 96]. The model can be found, for example, in [96] and references therein[3]. The building has eight floors, each having three degrees of freedom. The model is hence of dimension $n = 48$ and we overestimate the system order by $l = 50$. We simulate the model with a sampling time of $\delta t = 0.1$. The SDPs are solved with the Multi-Parametric Toolbox 3.0 [33] together with YALMIP [48]. The results presented below illustrate that the introduced method is simple to apply, can outperform model identification via standard tools and consecutive analysis of the identified model, and it provides good results in determining IQCs even for high dimensional systems and in the presence of measurement noise.

### 4.4.1 Determining dissipativity properties - a comparison to system identification methods

We measure one input-output trajectory of length $N = 2400$ with the output being subject to uniform multiplicative measurement noise of the form $\tilde{y}_k = (1 + \varepsilon_k)y_k$ with $\varepsilon_k \in [-\bar{\varepsilon}, \bar{\varepsilon}]$ at different noise levels $\bar{\varepsilon} > 0$ representing the SNR. The true system has an input feedforward passivity index of $\nu_{\text{tr}} = -0.001$ and, as also stated in [19], an $\mathcal{L}_2$-gain of $\gamma_{\text{tr}} = 0.0052$. Table 4.2 provides the estimated operator gain and passivity index via a simple bisection method together with Algorithm 1 for $L = 1050$ and $K = 3$. It can be seen that even for high levels of noise, the estimate of the operator gain as well as the passivity parameter are consistently close to the true values. Note that increasing $l$ leads to almost identical results, while after significantly decreasing $l$ the resulting values become more and more conservative, as to be expected and also further discussed in the next subsection.

For the same noise levels $\bar{\varepsilon} = 0.01, 0.10, 0.25, 0.50$, we now apply standard system identification tools, i.e., MATLAB functions *ssest* (estimates state-space model by initializing the parameter via a subspace approach or an iterative rational function

[3]The authors of [96] made their MATLAB files available on http://verivital.com/hyst/pass-order-reduction/.

**Table 4.2.** $\mathcal{L}_2$-gain estimates $\hat{\gamma}$ and input feedforward passivity index estimates $\hat{\nu}$ from noisy measurements via Algorithm 1 for increasing noise levels.

| $\bar{\varepsilon}$ | 0 | 1 % | 10 % | 25 % | 50 % |
|---|---|---|---|---|---|
| $\hat{\gamma}$ $(\cdot 10^{-3})$ | 5.2 | 5.2 | 5.1 | 5.1 | 5.0 |
| $\hat{\nu}$ $(\cdot 10^{-3})$ | -1.0 | -1.0 | -0.9 | -0.9 | -0.8 |

estimation approach and then refines the parameter values using the prediction error minimization approach), *ssregest* (estimates state-space model by reduction of a regularized ARX model), and *n4sid* (estimates state-space model using a subspace method). Each of the approaches is initialized with three different assumptions on the model order: $n_1 = 10$ (underestimated), $n_2 = 48$ (exact), $n_3 = 70$ (overestimated). After the model identification step, we determine the gain and the input feedforward passivity index with *norm(·,inf)* and *getPassiveIndex(·,'input')*, respectively. The result is summarized in Figure 4.4.

We can see that, in this high dimensional example, standard system identification tools from one noise-corrupted input-output trajectory produced quite variable results, which are also dependent on the assumed system order. The variance in the resulting estimates increases for higher noise levels even if the system order is known exactly. A plausible explanation for this is that the identification-based approach generally requires a two-step procedure of first determining a model and then performing model-based system analysis, where errors occurring in each step can be amplified. Please additionally note that the system identification function *ssest* required over one hour on an Intel i7 for a chosen system order of 70, while the computational expenses for the simple bisection method together with Algorithm 1 are below one minute. While a more in depth and especially theoretical comparison to system identification methods is part of future work, this one step approach to data-driven systems analysis provides a good basis for further improvements, especially in the case of tracking the noise and error propagation.

### 4.4.2 Determining an optimal IQC

From expert knowledge and insights to the mechanics of the building, one might have the suspicion that part of the system dynamics of the high dimensional system approximately behaves like a second order low-pass filter. Applying Theorem 4.3

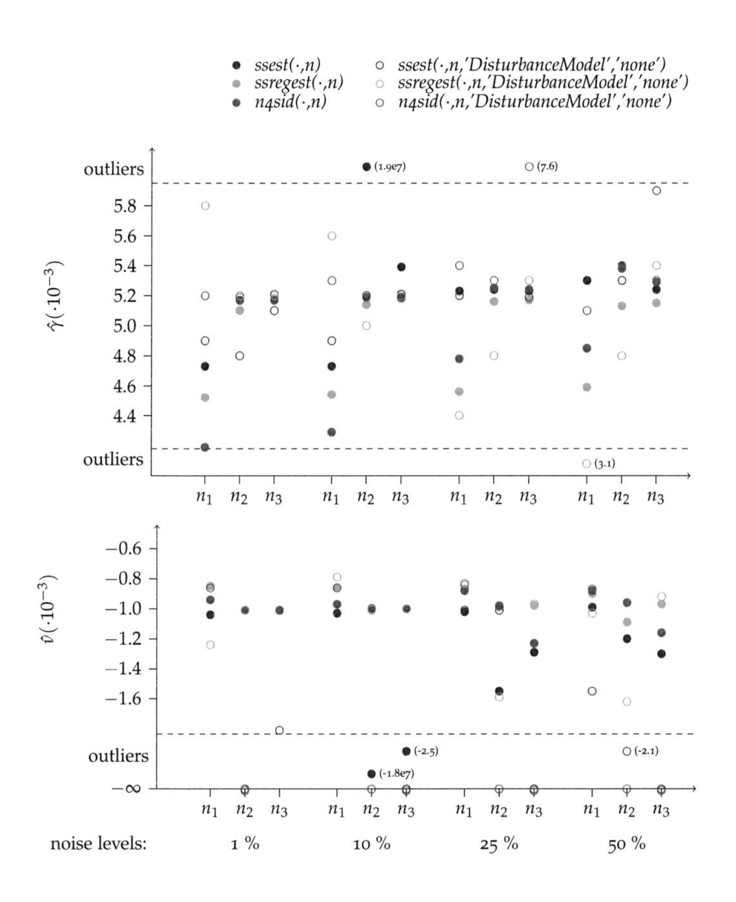


**Figure 4.4.** $\mathcal{L}_2$-gain estimates $\hat{\gamma}$ and input feedforward passivity index estimates $\hat{\nu}$ from noisy measurements with increasing noise levels, determined by applying different standard system identification methods for different given system orders $n_1 = 10$, $n_2 = 48$, $n_3 = 70$ and consecutive analysis methods.

along the lines of Remark 4.5, we can obtain a data-driven low-order model approximation with a guaranteed bound on the approximation error $\gamma_{\text{lo}}$. We select the IQC

$$M_\gamma = \begin{bmatrix} \gamma_{\text{lo}}^2 I_m & 0 \\ 0 & -I_p \end{bmatrix}, \quad \Psi(z) = \begin{bmatrix} I_m & 0 \\ -\Psi_{21}(z) & I_p \end{bmatrix},$$

with $\Psi_{21}(z) = \sum_{k=1}^{2} c_k^{(21)} B_k^{(21)}(z)$, where we choose two second order low-pass filters as basis functions. As a result, we hence find with $N = 1210$ that over the horizon $L - l = 500$ ($L = 550$), the tightest filter $\Psi_{21}(z)$, i.e., the best approximation $\mathcal{G}_{\text{lo}}$, with the chosen basis functions $B_k^{(21)}(z)$, $k = 1, 2$, is

$$\mathcal{G}_{\text{lo}} = \frac{2.67 \cdot 10^{-4}(10z + 1)}{z^2 + 0.5z + 0.1} + \frac{5.33 \cdot 10^{-5}(z + 1)}{z^2 - 1.2z + 0.7} \tag{4.16}$$

and the difference between the full-order model $\mathcal{G}_{\text{full}}$ and the lower-order approximation $\mathcal{G}_{\text{lo}}$ over the horizon $L - l = 500$ is bounded by $\gamma_{\text{lo}} = 0.0035$. As a reference value, we calculate with full model knowledge that the approximation error is indeed given by $\|\mathcal{G}_{\text{full}} - \mathcal{G}_{\text{lo}}\|_\infty = 0.0035$. Finding and verifying an IQC hence worked via simple computations.

In the following, we investigate how the choice of the horizon (cf. Table 4.3) as well as the choice of $l$ (cf. Table 4.4) influence the computed best approximation $\mathcal{G}_{\text{lo}}$ together with the resulting approximation error $\gamma_{\text{lo}}$.

Table 4.3 shows $\|\mathcal{G}_{\text{full}} - \mathcal{G}_{\text{lo}}\|_\infty$ for the computed best approximation $\mathcal{G}_{\text{lo}}$ together with the corresponding $\gamma_{\text{lo}}$ for the given $(L - l)$-IQC with different choices of $L$. We can see that shortening the horizon does not significantly deteriorate the computed $\mathcal{G}_{\text{lo}}$, and even for this high-dimensional system, $\gamma_{\text{lo}}$ quickly approaches the corresponding value of the infinite horizon for increasing $L$.

**Table 4.3.** $\|\mathcal{G}_{\text{full}} - \mathcal{G}_{\text{lo}}\|_\infty$, where $\mathcal{G}_{\text{lo}}$ and corresponding $\gamma_{\text{lo}}$ are computed from data for different choices of $L$ (i.e., over different horizons $L - l$).

| $L - l$ | 100 | 200 | 300 | 400 | 500 |
|---|---|---|---|---|---|
| $\|\mathcal{G}_{\text{full}} - \mathcal{G}_{\text{lo}}\|_\infty$ $(\cdot 10^{-3})$ | 3.7 | 3.6 | 3.5 | 3.5 | 3.5 |
| $\gamma_{\text{lo}}$ $(\cdot 10^{-3})$ | 2.9 | 3.3 | 3.4 | 3.4 | 3.5 |

The results in Table 4.4 show that there is no change in the result for different $l \geq n$, as implied by the theoretical analysis in Section 4.1. Furthermore, even for smaller $l$, we receive an upper bound on the true $(L-l)$-IQC, which might however be conservative (especially for $l \ll n$). This is well aligned with the theoretical results as explained in Section 4.1.

**Table 4.4.** $\|\mathcal{G}_{\text{full}} - \mathcal{G}_{\text{lo}}\|_\infty$, where $\mathcal{G}_{\text{lo}}$ and corresponding $\gamma_{\text{lo}}$ are computed from data for different assumed upper bounds on the lag of the system $l$.

| $l$ | 10 | 28 | 48 | 50 | 70 |
|---|---|---|---|---|---|
| $\|\mathcal{G}_{\text{full}} - \mathcal{G}_{\text{lo}}\|_\infty$ $(\cdot 10^{-3})$ | 5.2 | 4.3 | 3.5 | 3.5 | 3.5 |
| $\gamma_{\text{lo}}$ $(\cdot 10^{-3})$ | 65.2 | 4.4 | 3.5 | 3.5 | 3.5 |

Finally, we add measurement noise and calculate the approximation error $\gamma$ considering the IQC with $\mathcal{G}_{\text{lo}}$ as in (4.16) via Algorithm 1 and $K = 3$. We can see in Table 4.5 that even for high dimensional systems, verifying an IQC works remarkably well also in the presence of significant measurement noise.

**Table 4.5.** Approximation error $\gamma_{\text{lo}}$ computed from data via Algorithm 1 for different noise levels with $\mathcal{G}_{\text{lo}}$ from (4.16).

| $\bar{\varepsilon}$ | 0 % | 1 % | 5 % | 10 % | 25 % | 50 % |
|---|---|---|---|---|---|---|
| $\gamma$ $(\cdot 10^{-3})$ | 3.5 | 3.5 | 3.5 | 3.4 | 3.4 | 3.4 |

## 4.5 Summary

In this chapter, we introduced a simple approach to verify and find IQCs from only one input-output trajectory of an unknown LTI system. The developed approach is a one-step procedure examining the semidefiniteness of a single data-dependent matrix. We also provided guarantees to find the tightest system property descriptions via a simple SDP, as well as introduced results on the difference between system properties over the finite and the infinite horizon. A high-dimensional application example showed the potential of the approach even for challenging applications as well as in the presence of noise.

One of the advantages of the proposed method is certainly the simplicity of the verification approach: collecting data in a matrix and evaluating its definiteness. The only two requirements for its application are that the input trajectory of the available data is persistently exciting of sufficient order and that a (possibly conservative) upper bound on the lag of the system (or its order) is known. Compared to the iterative framework presented in Chapter 3, this method is more data-efficient and provides a unitary approach for general MIMO systems. Furthermore, the approach presented in this chapter can also be used to initialize an iterative approach to speed up its convergence.

The general concept developed in this chapter also offers a promising basis to extend the ideas to nonlinear systems with ideas from, e.g., [9, 46, AK25], where initial notions in the direction of Willems' fundamental lemma for (special classes of) nonlinear systems are introduced. Furthermore, the presented ideas can also be used for data-driven controller design with closed-loop dissipativity guarantees as introduced in [AK27].

Remaining open questions are tight results for the infinite horizon, and even more importantly, guarantees in the case of noisy data. While the presented one-step approach to verifying input-output system properties offers a good basis for tracking the noise and error propagation, we will approach this question from a different viewpoint using robust analysis and control tools in the next section.

# Chapter 5

# Bounds on dissipativity via a robust viewpoint

In most existing approaches to data-driven dissipativity analysis, including iterative approaches (cf. Chapter 3 and, e.g., [38, 72, 82, 106]) and one-shot methods based on Willems' fundamental lemma (cf. Chapter 4), sharp results are only available for system properties holding over the finite-time horizon. More importantly, none of the existing methods can provide deterministic guarantees on dissipativity properties from general noise-corrupted finite-length input-output trajectories. Therefore, we develop in this chapter an approach that certifies dissipativity properties of System (2.1) from one noise-corrupted trajectory of finite length with guarantees over the infinite horizon via a robust viewpoint. To this end, we consider two cases:

- **Input-state data** (Sections 5.1 and 5.2):

  We assume that $A$ and $B$ in (2.1) are unknown, but one input-state trajectory $\{u_k\}_{k=0}^{N-1}$, $\{x_k\}_{k=0}^{N}$ is available. Further, we assume that[1] $C$ and $D$ are known.

- **Input-output data** (Sections 5.3 and 5.4):

  We assume that $A$, $B$, $C$, and $D$ in (2.1) are unknown, but one input-output trajectory $\{u_k, y_k\}_{k=0}^{N-1}$ is available as well as an upper bound on the lag of the system $l \geq \underline{l}$.

For each case, we in turn distinguish between noise-free measurements of System (2.1), as considered in Section 5.1 and Section 5.3, and noisy data, as covered

[1] It is straightforward to extend the presented results to the case that $C$ and $D$ are unknown, but measurements of $\{y_k\}_{k=0}^{N-1}$ are available.

in Section 5.2 and Section 5.4. We collect the respective data sequences $\{u_k\}_{k=0}^{N-1}$, $\{x_k\}_{k=0}^{N}$, or $\{y_k\}_{k=0}^{N-1}$ in the following matrices

$$X = \begin{bmatrix} x_0 & x_1 & \cdots & x_{N-1} \end{bmatrix}, \qquad X_+ = \begin{bmatrix} x_1 & x_2 & \cdots & x_N \end{bmatrix},$$
$$U = \begin{bmatrix} u_0 & u_1 & \cdots & u_{N-1} \end{bmatrix}, \qquad Y = \begin{bmatrix} y_0 & y_1 & \cdots & y_{N-1} \end{bmatrix}.$$

On the basis of the data captured in these matrices, we develop a framework to guarantee dissipativity. More specifically, the high-level idea is to describe all systems which are consistent with these data matrices and then to apply tools from robust analysis and control to introduce computationally attractive methods to certify and determine dissipativity properties with guarantees, given that the measured trajectory is informative enough. One condition in this respect, which will play an important role in the following sections, is the rank condition

$$\operatorname{rank} \begin{bmatrix} X \\ U \end{bmatrix} = n + m. \tag{5.1}$$

Generally speaking, this condition can be ensured by requiring that the input of the measured trajectory is persistently exciting of a sufficiently high order (cf. Definition 2.3).

The remainder of this chapter is structured as follows. In Section 5.1, we introduce an equivalent dissipativity characterization purely on the basis of input-state data, followed by a noisy consideration thereof in Section 5.2, where we provide a nonconservative robust verification framework for dissipativity properties. Next, we generalize the results to input-output trajectories first in the noise-free case in Section 5.3, which we then extend to a dissipativity analysis approach from noise-corrupted input-output trajectories in Section 5.4. Finally, we apply the introduced approach to real-world data of a two-tank water system in Section 5.5 and end with a short summary in Section 5.6.

This chapter is based on and taken in parts literally from [AK5][2], [AK6][3].

[2]A. Koch, J. Berberich, F. Allgöwer. "Provably robust verification of dissipativity properties from data." In: *IEEE Trans. Automat. Control*, early access (2021). doi: 10.1109/TAC.2021.3116179 © 2021 IEEE.

[3]A. Koch, J. Berberich, F. Allgöwer. "Verifying dissipativity properties from noise-corrupted input-state data." In: *Proc. 59th IEEE Conf. on Decision and Control* (2020). pp. 616–621 © 2020 IEEE.

## 5.1 Dissipativity from input-state trajectories

With the standard results from the dissipativity literature as stated in Chapter 2 together with the data captured in the matrices $X$, $X_+$, and $U$, we can directly state an equivalent formulation for dissipativity from noise-free input-state trajectories. The resulting necessary and sufficient condition can be stated via a simple linear matrix inequality (LMI), which can be solved using standard solvers.

**Theorem 5.1.** *Suppose $\{u_k\}_{k=0}^{N-1}$, $\{x_k\}_{k=0}^{N}$ is an input-state trajectory of System (2.1).*

*(i) If there exists a $\mathcal{X} = \mathcal{X}^\top \succeq 0$ such that*

$$X_+^\top \mathcal{X} X_+ - X^\top \mathcal{X} X - \begin{bmatrix} U \\ CX + DU \end{bmatrix}^\top \begin{bmatrix} R & S^\top \\ S & Q \end{bmatrix} \begin{bmatrix} U \\ CX + DU \end{bmatrix} \preceq 0 \tag{5.2}$$

*holds and, additionally, the rank condition in (5.1) is satisfied, then System (2.1) is $(Q, S, R)$-dissipative.*

*(ii) If there exists no $\mathcal{X} = \mathcal{X}^\top \succeq 0$ such that (5.2) holds, then System (2.1) is not $(Q, S, R)$-dissipative.*

*Proof.* Substituting $X_+ = AX + BU$, the semidefiniteness condition in (5.2) can be equivalently written as

$$\begin{bmatrix} X \\ U \end{bmatrix}^\top \begin{bmatrix} A^\top \mathcal{X} A - \mathcal{X} - \hat{Q} & A^\top \mathcal{X} B - \hat{S} \\ (A^\top \mathcal{X} B - \hat{S})^\top & -\hat{R} + B^\top \mathcal{X} B \end{bmatrix} \begin{bmatrix} X \\ U \end{bmatrix} \preceq 0 \tag{5.3}$$

with $\hat{Q} = C^\top Q C$, $\hat{S} = C^\top S + C^\top Q D$, and $\hat{R} = D^\top Q D + (D^\top S + S^\top D) + R$.

(i) With (5.1), the semidefiniteness condition in (5.3) in turn implies that (2.6) holds, which implies dissipativity by Theorem 2.1.

(ii) If there exists no $\mathcal{X} = \mathcal{X}^\top \succeq 0$ such that (5.2) holds, this directly implies that (2.6) is not negative semidefinite for any $\mathcal{X}$, i.e., System (2.1) is not dissipative by Theorem 2.1.

■

**Remark 5.1.** The condition in (5.1) can easily be checked given the available data. By [108, Corollary 2], this rank condition can also be enforced by requiring or choosing the input $\{u_k\}_{k=0}^{N-1}$ to be persistently exciting of order $n+1$. Note that the latter condition requires a minimum length of the input trajectory, namely $N \geq (m+1)n + m$.

**Remark 5.2.** If $C$ and $D$ are unknown but measurements of the output are available instead, then one can equivalently substitute $Y = CX + DU$ in Equation (5.2) of the feasibility problem.

The result stated in Theorem 5.1 is conceptually similar to the approach in [103], where methods for data-based system analysis (e.g., controllability, stability) are provided by verifying such properties for all systems which are consistent with the data. The data-based formulation of dissipativity given by Theorem 5.1 is particularly simple and only requires solving a single SDP. The proof relies on the fact that, if the matrix $\begin{bmatrix} X \\ U \end{bmatrix}$ has full row rank, then it spans all possible input-state trajectories. Multiplying (2.6) from both sides by this matrix and exploiting the system dynamics $X_+ = AX + BU$, we obtain the stated result.

In contrast to the other approaches presented in Chapter 3 and Chapter 4, we exploited here the state-space definition of dissipativity, which takes a difference viewpoint, i.e., looks at the difference at two time points (cf. Definition 2.4). This yields the advantages that sharp results on the infinite horizon can be obtained as well as rigorous guarantees in the noisy case, as will be discussed in the next section.

## 5.2 Dissipativity from noisy input-state trajectories

While Section 5.1 provides a simple, computationally attractive condition to verify dissipativity properties of unknown systems, it assumes that exact measurements of input and state variables are available. This assumption does rarely hold in practice. Therefore, in this section, we extend the results to the case that the measured data are affected by noise. More precisely, we consider in this section a variation of (2.1) that is disturbed by process noise of the form

$$\begin{aligned} x_{k+1} &= Ax_k + Bu_k + B_w w_k, \\ y_k &= Cx_k + Du_k, \end{aligned} \tag{5.4}$$

where $w_k \in \mathbb{R}^{m_w}$ denotes the noise and $B_w \in \mathbb{R}^{n \times m_w}$ is a known matrix with full column rank describing the influence of the noise on the system dynamics. Note that we refer to $w_k$ as (process) noise throughout this chapter. Alternatively, $w_k$ can also be interpreted as an input disturbance, possibly even capturing slight nonlinearities.

**Remark 5.3.** Note that including a known matrix $B_w$ into the consideration offers the possibility to include additional knowledge on the influence of the process noise on the system into the analysis, which can improve the results. If no additional information on the effect of the noise on the different states is available, one can simply choose the identity matrix $B_w = I$.

We denote the actual noise sequence which yields the available input-state trajectory $\{u_k\}_{k=0}^{N-1}$, $\{x_k\}_{k=0}^{N}$ by $\{\hat{w}_k\}_{k=0}^{N-1}$. While this noise sequence $\{\hat{w}_k\}_{k=0}^{N-1}$ is unknown, we assume that some information on the noise is available in form of a bound on the stacked matrix

$$\hat{W} = \begin{bmatrix} \hat{w}_0 & \hat{w}_1 & \cdots & \hat{w}_{N-1} \end{bmatrix}$$

as specified in the following assumption.

**Assumption 5.1.** *The matrix $\hat{W}$ denoting the stacked process noise $\{\hat{w}_k\}_{k=0}^{N-1}$ is an element of the set*

$$\mathcal{W} = \{W \in \mathbb{R}^{m_w \times N} | \begin{bmatrix} W^\top \\ I \end{bmatrix}^\top \begin{bmatrix} Q_w & S_w \\ S_w^\top & R_w \end{bmatrix} \begin{bmatrix} W^\top \\ I \end{bmatrix} \succeq 0\} \tag{5.5}$$

*where $Q_w \in \mathbb{R}^{N \times N}$, $S_w \in \mathbb{R}^{N \times m_w}$, and $R_w \in \mathbb{R}^{m_w \times m_w}$, with $Q_w \prec 0$.*

This quadratic bound on the noise matrix $\hat{W}$ is a flexible noise or disturbance description. Similar bounds on the noise were also used, for example, in [10, 102, AK3]. As discussed in more detail in these references, this quadratic matrix bound can incorporate, for example, bounds on sequences ($\|\hat{w}\| \leq \bar{w}_s$) or the maximal singular value of $\hat{W}$. Furthermore, it can also capture bounds on separate components ($\|\hat{w}_k\| \leq \bar{w}$ for all $k$), i.e., point-wise bounds, however, only by introducing conservatism (see, e.g., [10, 102] for details).

Due to the presence of noise, there generally exist multiple matrix pairs $(A_d, B_d)$ which are consistent with the data for some noise sequence $W \in \mathcal{W}$. The set of all such matrix pairs consistent with the input-state data and the noise bound is in the following denoted by

$$\Sigma_{X,U} = \{(A_\mathrm{d}, B_\mathrm{d}) | X_+ = A_\mathrm{d}X + B_\mathrm{d}U + B_w W, W \in \mathcal{W}\}.$$

By assumption, this set includes the system matrices $(A, B)$ which generated the data.

To verify that System (2.1) is indeed $(Q, S, R)$-dissipative from noisy data, it is necessary to verify that *all* systems that are consistent with the data are $(Q, S, R)$-dissipative. Therefore, we first develop a data-driven open-loop representation of all systems that are consistent with the available input-state trajectory. This approach is inspired by ideas and results in [75, AK3], where the closed loop is parameterized on the basis of an input-state trajectory in order to design (robust) controllers.

**Lemma 5.1.** *If there exists a matrix $\mathcal{K}$ such that*

$$\begin{bmatrix} X \\ U \end{bmatrix} \mathcal{K} = I, \tag{5.6}$$

*then $\Sigma_{X,U}$ is equal to the set of $(A_\mathrm{d}, B_\mathrm{d})$ satisfying*

$$\begin{bmatrix} A_\mathrm{d} & B_\mathrm{d} \end{bmatrix} = (X_+ - B_w W)\mathcal{K} \tag{5.7}$$

*for some $W \in \mathcal{W}$ with*

$$(X_+ - B_w W) \begin{bmatrix} X \\ U \end{bmatrix}^\perp = 0. \tag{5.8}$$

*Proof.* First note that, as explained in [AK3, Theorem 4], the constraint in (5.8) is, by the Fredholm alternative, equivalent to the existence of a solution $\mathtt{V}$ to the system of linear equations

$$\mathtt{V} \begin{bmatrix} X \\ U \end{bmatrix} = X_+ - B_w W. \tag{5.9}$$

(i) Assume (5.7) holds for some $W \in \mathcal{W}$ with (5.8). We need to show that there exist $(\bar{A}, \bar{B})$, $\bar{W} \in \mathcal{W}$ such that $\begin{bmatrix} \bar{A} & \bar{B} \end{bmatrix} = (X_+ - B_w W)\mathcal{K}$ with $X_+ = \bar{A}X + \bar{B}U +$

$B_w\bar{W}$. We know that for all $W \in \mathcal{W}$ satisfying (5.8), there exists a solution $\mathtt{V}$ to (5.9). Hence the choice $\begin{bmatrix} \bar{A} & \bar{B} \end{bmatrix} = \mathtt{V}$ from (5.9) ensures $X_+ = \bar{A}X + \bar{B}U + B_w\bar{W}$ with $\bar{W} = W$, and $\begin{bmatrix} \bar{A} & \bar{B} \end{bmatrix} = \mathtt{V}$ also satisfies

$$\begin{bmatrix} \bar{A} & \bar{B} \end{bmatrix} = \begin{bmatrix} \bar{A} & \bar{B} \end{bmatrix} \begin{bmatrix} X \\ U \end{bmatrix} \mathcal{K} = (X_+ - B_w W)\mathcal{K}.$$

(ii) For any $(A_{\mathrm{d}}, B_{\mathrm{d}}) \in \Sigma_{X,U}$, there exists per definition a $W \in \mathcal{W}$ such that $A_{\mathrm{d}}X + B_{\mathrm{d}}U = X_+ - B_w W$. Hence, there exists a solution $\mathtt{V}$ to (5.9), which implies that (5.8) holds. Multiplying $\mathcal{K}$ from the right on both sides yields

$$\begin{bmatrix} A_{\mathrm{d}} & B_{\mathrm{d}} \end{bmatrix} \begin{bmatrix} X \\ U \end{bmatrix} \mathcal{K} = \begin{bmatrix} A_{\mathrm{d}} & B_{\mathrm{d}} \end{bmatrix} = (X_+ - B_w W)\mathcal{K}.$$

■

Lemma 5.1 provides an equivalent description of $\Sigma_{X,U}$, and hence, a simple parameterization of all systems consistent with the available input-state trajectory. For some applications of this parameterization, it is computationally advantageous to consider the following superset of $\Sigma_{X,U}$. Let $\Sigma^{\mathrm{s}}_{X,U}$ denote the set of systems which are described by

$$\begin{bmatrix} A^{\mathrm{s}}_{\mathrm{d}} & B^{\mathrm{s}}_{\mathrm{d}} \end{bmatrix} = (X_+ - B_w W)\mathcal{K} \quad \text{for some } W \in \mathcal{W}. \tag{5.10}$$

We hence drop the condition in (5.8), which immediately shows $\Sigma_{X,U} \subseteq \Sigma^{\mathrm{s}}_{X,U}$.

The simple characterization of all systems consistent with the data in Lemma 5.1 and the superset described in (5.10) prove to be very valuable for different data-driven analysis objectives. For data-driven reachability analysis and set-based estimation, for example, the application of zonotopes and constraint zonotopes allows for the representation of the sets $\Sigma^{\mathrm{s}}_{X,U}$ and $\Sigma_{X,U}$, respectively, in a computationally attractive way. Exploiting these set representations in reachability and estimation methods yields data-driven reachability analysis and estimation results with deterministic containment guarantees [11, AK1, AK2]. Furthermore, the parameterization in Lemma 5.1 also allows to infer robust bounds on dissipativity in the case of quadratic bounds on the noise matrix $\hat{W}$ by applying robust systems analysis tools. This leads to easily verifiable conditions to guarantee system properties from noisy data of finite length as introduced in [AK6, Theorem 9], based on the simple characterization $\Sigma^{\mathrm{s}}_{X,U}$ containing all system matrices consistent with the data.

However, the result in [AK6, Theorem 9] yet has two disadvantages. Firstly, to obtain a simple SDP for verifying dissipativity in [AK6, Theorem 9], we can only consider the superset $\Sigma^{\mathrm{s}}_{X,U}$ possibly introducing conservatism, and secondly, the computational expenses for verifying or determining dissipativity properties increase with the data length. To develop nonconservative and computationally favorable conditions for dissipativity, we make use of another equivalent representation of the set $\Sigma_{X,U}$ recently introduced in [102].

**Lemma 5.2.** *It holds that*

$$\Sigma_{X,U} = \{(A_{\mathrm{d}}, B_{\mathrm{d}}) | \begin{bmatrix} A_{\mathrm{d}}^\top \\ B_{\mathrm{d}}^\top \\ I \end{bmatrix}^\top \begin{bmatrix} \bar{Q}_w & \bar{S}_w \\ \bar{S}_w^\top & \bar{R}_w \end{bmatrix} \begin{bmatrix} A_{\mathrm{d}}^\top \\ B_{\mathrm{d}}^\top \\ I \end{bmatrix} \succeq 0\} \tag{5.11}$$

*with*

$$\bar{Q}_w = \begin{bmatrix} X \\ U \end{bmatrix} Q_w \begin{bmatrix} X \\ U \end{bmatrix}^\top, \quad \bar{S}_w = -\begin{bmatrix} X \\ U \end{bmatrix} (Q_w X_+^\top + S_w B_w^\top),$$
$$\bar{R}_w = X_+ Q_w X_+^\top + X_+ S_w B_w^\top + B_w S_w^\top X_+^\top + B_w R_w B_w^\top.$$

*Proof.* This statement follows from [102, Lemma 4 and Remark 2]. ∎

Using the adapted parameterization of $\Sigma_{X,U}$ in Lemma 5.2, we rewrite the problem of robustly verifying dissipativity properties from noisy data in such a form that we can apply robust analysis tools from the literature. To this end, we start by writing the set of all LTI systems consistent with the data as

$$x_{k+1} = \begin{bmatrix} A_{\mathrm{d}} & B_{\mathrm{d}} \end{bmatrix} \begin{bmatrix} x_k \\ u_k \end{bmatrix}$$

for some $(A_{\mathrm{d}}, B_{\mathrm{d}}) \in \Sigma_{X,U}$. We can equivalently reformulate this uncertain system as a linear fractional transformation (LFT) [114] of a nominal system with the 'uncertainty' $\begin{bmatrix} A_{\mathrm{d}} & B_{\mathrm{d}} \end{bmatrix}$, i.e.,

$$\begin{bmatrix} x_{k+1} \\ \tilde{z}_k \end{bmatrix} = \begin{bmatrix} 0 & I \\ I & 0 \end{bmatrix} \begin{bmatrix} \begin{bmatrix} x_k \\ u_k \end{bmatrix} \\ \tilde{w}_k \end{bmatrix}, \quad \tilde{w}_k = \begin{bmatrix} A_{\mathrm{d}} & B_{\mathrm{d}} \end{bmatrix} \tilde{z}_k, \tag{5.12}$$

with $(A_\mathrm{d}, B_\mathrm{d}) \in \Sigma_{X,U}$. Applying some additional tools from the robust analysis and control literature [86, 87] allows to introduce nonconservative and computationally attractive dissipativity conditions. To this end, we define

$$\begin{bmatrix} \tilde{R} & \tilde{S}^\top \\ \tilde{S} & \tilde{Q} \end{bmatrix} = \begin{bmatrix} R & S^\top \\ S & Q \end{bmatrix}^{-1}, \tag{5.13}$$

assuming that the inverse exists.

**Theorem 5.2.** *Let $\tilde{R} \succeq 0$. If there exist a matrix $\mathcal{X} = \mathcal{X}^\top \succ 0$ and a scalar $\tau > 0$ such that*

$$\begin{bmatrix} \star & \star & \star \\ \star & \star & \star \\ \hline \star & \star & \star \\ \star & \star & \star \\ \hline \star & \star & \star \\ \star & \star & \star \end{bmatrix}^\top \left[\begin{array}{cc|cc|cc} -\mathcal{X} & 0 & 0 & 0 & 0 & 0 \\ 0 & \mathcal{X} & 0 & 0 & 0 & 0 \\ \hline 0 & 0 & -\tilde{R} & -\tilde{S}^\top & 0 & 0 \\ 0 & 0 & -\tilde{S} & -\tilde{Q} & 0 & 0 \\ \hline 0 & 0 & 0 & 0 & -\tau \bar{Q}_w & -\tau \bar{S}_w \\ 0 & 0 & 0 & 0 & -\tau \bar{S}_w^\top & -\tau \bar{R}_w \end{array}\right] \begin{bmatrix} (I \;\; 0) & 0 & C^\top \\ 0 & -I & 0 \\ \hline (0 \;\; I) & 0 & D^\top \\ 0 & 0 & -I \\ \hline I & 0 & 0 \\ 0 & I & 0 \end{bmatrix} \succ 0 \tag{5.14}$$

*holds, then System (2.1) is $(Q, S, R)$-dissipative for all matrices consistent with the data $(A_\mathrm{d}, B_\mathrm{d}) \in \Sigma_{X,U}$.*

*Proof.* By the full-block S-procedure [86] and using Lemma 5.2, (5.14) implies that

$$\begin{bmatrix} \star & \star \\ \star & \star \\ \hline \star & \star \\ \star & \star \end{bmatrix}^\top \left[\begin{array}{cc|cc} -\mathcal{X} & 0 & 0 & 0 \\ 0 & \mathcal{X} & 0 & 0 \\ \hline 0 & 0 & -\tilde{R} & -\tilde{S}^\top \\ 0 & 0 & -\tilde{S} & -\tilde{Q} \end{array}\right] \begin{bmatrix} A_\mathrm{d}^\top & C^\top \\ -I & 0 \\ \hline B_\mathrm{d}^\top & D^\top \\ 0 & -I \end{bmatrix} \succ 0$$

holds for all $(A_\mathrm{d}, B_\mathrm{d}) \in \Sigma_{X,U}$. Using the dualization lemma [87], this in turn proves that

$$\begin{bmatrix} \star & \star \\ \star & \star \\ \hline \star & \star \\ \star & \star \end{bmatrix}^\top \left[\begin{array}{cc|cc} -\mathcal{X}^{-1} & 0 & 0 & 0 \\ 0 & \mathcal{X}^{-1} & 0 & 0 \\ \hline 0 & 0 & -R & -S^\top \\ 0 & 0 & -S & -Q \end{array}\right] \begin{bmatrix} I & 0 \\ A_\mathrm{d} & B_\mathrm{d} \\ \hline 0 & I \\ C & D \end{bmatrix} \prec 0$$

holds for all $(A_\mathrm{d}, B_\mathrm{d}) \in \Sigma_{X,U}$.

By Theorem 2.1, this implies that System (2.1) is $(Q, S, R)$-dissipative for all matrices consistent with the data $(A_{\mathrm{d}}, B_{\mathrm{d}}) \in \Sigma_{X,U}$, which concludes the proof. ■

Theorem 5.2 provides a powerful tool to verify dissipativity via a simple LMI (5.14) based on noisy input-state measurements captured in the matrices $\bar{Q}_w$, $\bar{S}_w$, and $\bar{R}_w$. Note that the number of decision variables does not grow with the data length. With (5.14) being linear in the matrices $(\tilde{Q}, \tilde{S}, \tilde{R})$, optimizing over specific dissipativity related parameters can be done via a simple SDP. For finding the $\mathcal{L}_2$-gain $\gamma$, e.g., choose $\tilde{R} = \frac{1}{\gamma^2} I$, $\tilde{S} = 0$, $\tilde{Q} = -I$ and minimize $-\frac{1}{\gamma^2}$ such that (5.14) holds.

Alternatively, it is possible to derive an analogous result to Theorem 5.2 by first applying the dualization lemma directly to the parameterization in (5.11) and then apply the S-procedure[4]. However, this approach can potentially have a numerical disadvantage as, instead of the inverse in (5.13), an inverse of the data-dependent matrix $\begin{bmatrix} \bar{Q}_w & \bar{S}_w \\ \bar{S}_w^\top & \bar{R}_w \end{bmatrix}$ needs to be computed.

**Remark 5.4.** Theorem 5.2 reduces the conservatism compared to [AK6, Theorem 9] by not considering a superset of system matrices consistent with the data but an equivalent representation (cf. Lemma 5.2). Furthermore, the result in Theorem 5.2 is nonconservative in the sense that feasibility of (5.14) is not only sufficient but also necessary for (2.6) to hold for all $(A_{\mathrm{d}}, B_{\mathrm{d}}) \in \Sigma_{X,U}$, i.e., the condition is equivalent to dissipativity of all systems in $\Sigma_{X,U}$ with a common quadratic storage function. This is due to the fact that, for one quadratic constraint, the S-procedure is both necessary and sufficient [102]. However, verifying dissipativity on the basis of Theorem 5.2 introduces some conservatism, as detailed in the following. Most significantly, we only consider a common storage function for all systems consistent with the data, which is a simplification often applied in robust analysis and controller design tools (cf. [87]). Reducing the conservatism by considering parameter-dependent storage functions is an interesting issue for future research. Furthermore, we need to restrict our attention to $\mathcal{X} \succ 0$ in Theorem 5.2 due to the dualization step, which, however, only yields a marginal difference to $\mathcal{X} \succeq 0$ (cf. Theorem 2.1) in any numerical implementation in practice. Finally, it is possible to further reduce conservatism by including more precise noise bounds or prior knowledge on the mathematical model using similar tools as in [10].

[4]This dual result together with its derivation can be found in an earlier version of [AK7]: https://arxiv.org/pdf/2006.05974v1.pdf, Theorem 12.

**Remark 5.5.** An alternative to direct data-driven systems analysis as presented in Theorem 5.2 is to consecutively identify a mathematical model from data and analyze it using model-based tools. However, most existing system identification methods provide only asymptotic guarantees in the case of noisy data. Nonasymptotic guarantees in identifying a system from data of finite length is ongoing research. Interesting results in this direction include [56], where probabilistic assumptions on the noise yield probabilistic guarantees on the identified model. Set membership estimation is another very related approach considering deterministic noise bounds, where nonconservative but computationally tractable error bounds are a key challenge [61]. In comparison, the presented approach yields computationally attractive LMI conditions independent of the data length for verifying dissipativity with guarantees under deterministic process noise bounds.

**Example 5.1.** To illustrate the introduced approach, we apply it to a numerical example. We choose a randomly generated system of order $n = 5$ with two inputs and outputs $m = p = 2$. We simulate a trajectory with $u_k$, $k = 0, \dots, N-1$, uniformly sampled in $[-1, 1]$ for different lengths $N$, and we sample the disturbance $\hat{w}_k$ uniformly from the ball $\|\hat{w}_k\| \leq \bar{w}$ for all $k = 0, \dots, N-1$, where $\bar{w} = 0.01$. This implies a bound on the measurement noise given by $\hat{W}\hat{W}^\top \preceq \bar{w}^2 N I$ for the respective data length $N$. The true $\mathcal{L}_2$-gain of the randomly generated system is $\gamma_{\mathrm{tr}} = 11.44$. Applying the LMI condition in (5.14) from Theorem 5.2, we retrieve an upper bound on the $\mathcal{L}_2$-gain that is guaranteed for all systems consistent with the data. The resulting upper bounds on the $\mathcal{L}_2$-gain for different data lengths $N$ are depicted in Figure 5.1.

Next, we take the same example and increase the noise level $\bar{w}$ from 0.001 to 0.02 for $N = 50$ data points each and again apply the LMI condition in (5.14) from Theorem 5.2. The resulting upper bounds on the $\mathcal{L}_2$-gain are depicted in Figure 5.2.

The results in Figure 5.1 and Figure 5.2 are generally well aligned with the theoretical guarantees. The computed $\hat{\gamma}$ is indeed always an upper bound on the true $\mathcal{L}_2$-gain $\gamma_{\mathrm{tr}}$. Furthermore, it can be seen that, for increasing noise bounds, the result becomes more conservative, as can be expected. More data points, i.e., longer trajectories, on the other hand, generally improve the result in this example. Note that this cannot be guaranteed in general since the quadratic bound on the noise does not tightly express point-wise bounds on the noise, as chosen in this example setup.

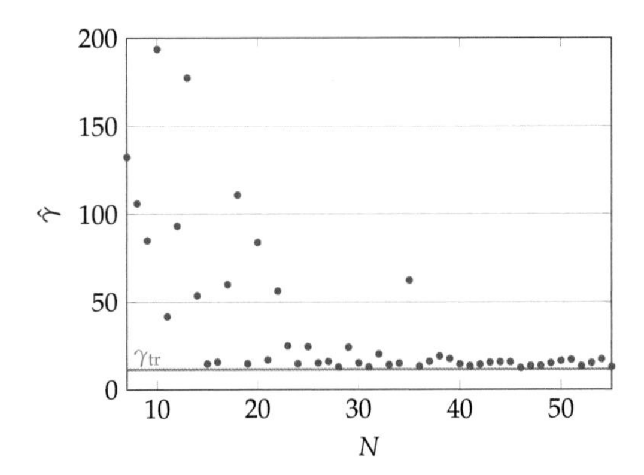


**Figure 5.1.** Guaranteed upper bounds on the $\mathcal{L}_2$-gain of the system in Example 5.1 from noisy input-state trajectories of different lengths $N$.

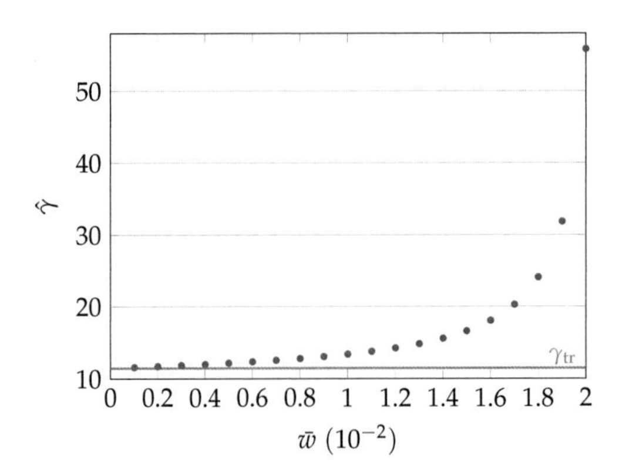


**Figure 5.2.** Guaranteed upper bounds on the $\mathcal{L}_2$-gain of the system in Example 5.1 from noisy input-state trajectories for increasing noise levels $\bar{w}$ and $N = 50$ data points.

As discussed above, the result in Theorem 5.2 is powerful, being nonconservative and computationally simple. However, it also requires the availability of state measurements, which can be restrictive in practice, where often only input-output data are available. In the next section, we therefore extend the results of Section 5.1 to an input-output setting.

## 5.3 Dissipativity from input-output trajectories

Instead of input-state measurements, we consider in this section the case where only an input-output sequence $\{u_k, y_k\}_{k=0}^{N-1}$ of System (2.1) is available to verify dissipativity properties. We use an extended state, based on $l$ consecutive inputs and outputs, in order to verify dissipativity properties, where $l$ is an upper bound on the lag $\underline{l}$ of the system (cf. Definition 2.1). The following lemma shows that this is in principle possible since the corresponding stacked system has the same input-output behavior as System (2.1).

**Lemma 5.3.** *Let $l \geq \underline{l}$. Then there exists an LTI system with the system matrices $\widetilde{A}, \widetilde{B}, \widetilde{C}, \widetilde{D}$ which can explain the data $\{u_k, y_k\}_{k=0}^{N-1}$, i.e., there exists $\xi_0$ such that for $k = 0, \dots, N-1$,*

$$\xi_{k+1} = \widetilde{A}\xi_k + \widetilde{B}u_k, \quad y_k = \widetilde{C}\xi_k + \widetilde{D}u_k, \tag{5.15}$$

*where the extended state is defined by*

$$\xi_k = \begin{bmatrix} u_{k-l}^\top & u_{k-l+1}^\top & \cdots & u_{k-1}^\top & y_{k-l}^\top & y_{k-l+1}^\top & \cdots & y_{k-1}^\top \end{bmatrix}^\top.$$

While Lemma 5.3 is a well-known fact, we nevertheless add a short proof for completeness and to provide some intuition.

*Proof.* The input-output behavior of System (2.1) over $l$ steps can be written as

$$\begin{bmatrix} y_{k-l} \\ y_{k-l+1} \\ \vdots \\ y_{k-1} \end{bmatrix} = \underbrace{\begin{bmatrix} C \\ CA \\ \vdots \\ CA^{l-1} \end{bmatrix}}_{\mathcal{O}_l} x_{k-l} + \underbrace{\begin{bmatrix} D & 0 & & \dots & 0 & 0 \\ CB & D & & \dots & 0 & 0 \\ \vdots & \ddots & \ddots & \ddots & \ddots & \vdots \\ CA^{l-2}B & CA^{l-3}B & \dots & CAB & CB & D \end{bmatrix}}_{R} \begin{bmatrix} u_{k-l} \\ u_{k-l+1} \\ \vdots \\ u_{k-1} \end{bmatrix}, \tag{5.16}$$

which yields with the introduced matrix notation

$$\begin{bmatrix} -R & I \end{bmatrix} \xi_k = \mathcal{O}_l x_{k-l}.$$

Using the definition of the lag, we know that $\mathcal{O}_l$ has full column rank, and hence, there exists a left-inverse $\mathcal{O}_l^{-1}$ (which has full row rank) such that

$$\underbrace{\mathcal{O}_l^{-1} \begin{bmatrix} -R & I \end{bmatrix}}_{T} \xi_k = x_{k-l}. \tag{5.17}$$

The general system description from (2.1) yields

$$y_k = CA^l x_{k-l} + \begin{bmatrix} CA^{l-1}B & \dots & CB \end{bmatrix} \begin{bmatrix} u_{k-l} \\ \vdots \\ u_{k-1} \end{bmatrix} + Du_k.$$

Together with $T$ as defined in (5.17), this leads to

$$\underbrace{\begin{bmatrix} u_{k-l+1} \\ \vdots \\ u_{k-1} \\ u_k \\ y_{k-l+1} \\ \vdots \\ y_{k-1} \\ y_k \end{bmatrix}}_{\xi_{k+1}} = \underbrace{\left( \begin{bmatrix} 0 & I & \dots & 0 & 0 & 0 & \dots & 0 \\ \vdots & \ddots & \ddots & \ddots & \vdots & & \ddots & \vdots \\ 0 & 0 & \dots & I & 0 & 0 & \dots & 0 \\ 0 & 0 & \dots & 0 & 0 & 0 & \dots & 0 \\ 0 & 0 & \dots & 0 & 0 & I & \dots & 0 \\ \vdots & \ddots & \ddots & \vdots & \vdots & & \ddots & \vdots \\ 0 & 0 & \dots & 0 & 0 & 0 & \dots & I \\ CA^{l-1}B & \dots & \dots & CB & 0 & 0 & \dots & 0 \end{bmatrix} + \begin{bmatrix} 0 \\ \vdots \\ 0 \\ 0 \\ 0 \\ \vdots \\ 0 \\ CA^l T \end{bmatrix} \right)}_{\tilde{A}} \underbrace{\begin{bmatrix} u_{k-l} \\ u_{k-l+1} \\ \vdots \\ u_{k-1} \\ y_{k-l} \\ y_{k-l+1} \\ \vdots \\ y_{k-1} \end{bmatrix}}_{\xi_k} + \underbrace{\begin{bmatrix} 0 \\ \vdots \\ 0 \\ I \\ 0 \\ \vdots \\ 0 \\ D \end{bmatrix}}_{\tilde{B}} u_k$$

$$y_k = \underbrace{\begin{bmatrix} 0 & \dots & 0 & I \end{bmatrix} \tilde{A}}_{\tilde{C}} \xi_k + \underbrace{D}_{\tilde{D}} u_k. \tag{5.18}$$

This proves that $(\tilde{A}, \tilde{B}, \tilde{C}, \tilde{D})$ can explain the input-output trajectory. ■

The converse of Lemma 5.3 follows trivially from its proof and the constructed extended system in (5.18): All input-output trajectories of the extended system in (5.18) with zero initial condition $\xi_0 = 0$ (or $\xi_0 \in \mathcal{X}_\xi$, where $\mathcal{X}_\xi$ denotes the set of reachable states) are also input-output trajectories of System (2.1). We can hence conclude that the extended system in (5.15) has the same input-output behavior as System (2.1) if the initial condition $\xi_0$ is restricted to the set of reachable states. This implies that both systems have the same input-output behavior for zero initial condition $\xi_0 = 0, x_0 = 0$. This allows to infer dissipativity of System (2.1) by verifying a dissipativity condition as given in (2.4) or in (2.7) for a potentially nonminimal system with the same input-output behavior. Hence, using Lemma 5.3, we can reduce the problem of verifying dissipativity from an input-output trajectory to the problem of verifying dissipativity from an input-state trajectory of System (5.15).

**Remark 5.6.** In a purely data-driven setup, knowledge on the lag $\underline{l}$ is often not available. Similarly to the results in Chapter 4, it is however sufficient for the results of this section to have an upper bound on $l$, as shown above.

Similar to the results of this section, [75] uses an extended state to design data-driven controllers but, instead of the lag $\underline{l}$, the system order $n$ is used. For MIMO systems, this can result in significantly larger state dimensions. Furthermore, [75] assumes controllability of the extended system, which is generally not the case unless $l{=}n$ and SISO systems are considered. Considering the more general case of uncontrollable extended systems will lead in the following to the challenge that the matrix containing the extended state does usually not have full row rank (cf. Condition (5.1)) even if the input is persistently exciting. However, applying Willems' fundamental lemma, as stated in Theorem 4.1, allows us to recover desirable guarantees.

Note that Willems' fundamental lemma in Theorem 4.1 requires controllability, which is not generally the case for the extended system in (5.15). However, since System (2.1) that generated the data is assumed to be minimal, we can apply Theorem 4.1 to describe all input-output trajectories of System (2.1) and hence also of the extended system for $\xi_0 \in \mathcal{X}_\xi$ applying Lemma 5.3. Using Lemma 5.3 and Theorem 4.1, we can hence determine dissipativity properties from input-output trajectories. For this, we collect the extended state data analogously to Section 5.1 in the following matrices

$$\Xi = \begin{bmatrix} \xi_l & \xi_{l+1} & \cdots & \xi_{N-1} \end{bmatrix}, \qquad \Xi_+ = \begin{bmatrix} \xi_{l+1} & \xi_{l+2} & \cdots & \xi_N \end{bmatrix},$$
$$Y_\Xi = \begin{bmatrix} y_l & y_{l+1} & \cdots & y_{N-1} \end{bmatrix}, \qquad U_\Xi = \begin{bmatrix} u_l & u_{l+1} & \cdots & u_{N-1} \end{bmatrix},$$

which directly leads us to the main result of this section.

**Theorem 5.3.** *Suppose $\{u_k, y_k\}_{k=0}^{N-1}$ is an input-output trajectory of System (2.1), which has lag $\underline{l}$. Let $l \geq \underline{l}$.*

*(i) If there exists a $\mathcal{X} = \mathcal{X}^\top \succeq 0$ such that*

$$\Xi_+^\top \mathcal{X} \Xi_+ - \Xi^\top \mathcal{X} \Xi - Y_\Xi^\top Q Y_\Xi - Y_\Xi^\top S U_\Xi - (S U_\Xi)^\top Y_\Xi - U_\Xi^\top R U_\Xi \preceq 0 \qquad (5.19)$$

*holds and, additionally, $\{u_k\}_{k=0}^{N-1}$ is persistently exciting of order $n + l + 1$, then System (2.1) is $(Q, S, R)$-dissipative.*

*(ii) If there exists no $\mathcal{X} = \mathcal{X}^\top \succeq 0$ such that (5.19) holds, then System (2.1) is not $(Q, S, R)$-dissipative.*

*Proof.*

(i) First, we notice that the data matrix $\Xi$ can be written as

$$\Xi = \begin{bmatrix} H_l(\{u_k\}_{k=0}^{N-2}) \\ H_l(\{y_k\}_{k=0}^{N-2}) \end{bmatrix} = \begin{bmatrix} u_0 & u_1 & \cdots & u_{N-l-1} \\ u_1 & u_2 & \cdots & u_{N-l} \\ \vdots & & & \vdots \\ u_{l-1} & u_l & \cdots & u_{N-2} \\ y_0 & y_1 & \cdots & y_{N-l-1} \\ y_1 & y_2 & \cdots & y_{N-l} \\ \vdots & & & \vdots \\ y_{l-1} & y_l & \cdots & y_{N-2} \end{bmatrix}.$$

Since there exists a controllable realization (of order $n$) with the same input-output behavior as the extended system, we can apply Willems' fundamental lemma. More specifically, if $\{u_k\}_{k=0}^{N-1}$ is persistently exciting of order $n + l$, then Willems' fundamental lemma stated in Theorem 4.1 guarantees that the columns in $\Xi$ span all possible input-output trajectories of System (2.1), and hence, the whole reachable state space of the extended system in (5.15). If $\{u_k\}_{k=0}^{N-1}$ is persistently exciting of order $n + l + 1$, then it additionally holds that $\begin{bmatrix} \Xi \\ U_\Xi \end{bmatrix}$ spans the space of all input-state trajectories of (5.15).

Using $\Xi_+ = \tilde{A}\Xi + \tilde{B}U_\Xi$ and rearranging (5.19), we obtain

$$\begin{bmatrix} \Xi \\ U_\Xi \end{bmatrix}^\top \begin{bmatrix} \widetilde{A}^\top \mathcal{X}\widetilde{A} - \mathcal{X} - \hat{Q} & \widetilde{A}^\top \mathcal{X}\widetilde{B} - \hat{S} \\ (\widetilde{A}^\top \mathcal{X}\widetilde{B} - \hat{S})^\top & -\hat{R} + \widetilde{B}^\top \mathcal{X}\widetilde{B} \end{bmatrix} \begin{bmatrix} \Xi \\ U_\Xi \end{bmatrix} \preceq 0, \tag{5.20}$$

with $\hat{Q}$, $\hat{S}$, $\hat{R}$ similar as in (2.6). Since $\begin{bmatrix} \Xi \\ U_\Xi \end{bmatrix}$ spans the space of all input-state trajectories, this implies

$$\begin{bmatrix} \xi_k \\ u_k \end{bmatrix}^\top \begin{bmatrix} \widetilde{A}^\top \mathcal{X}\widetilde{A} - \mathcal{X} - \hat{Q} & \widetilde{A}^\top \mathcal{X}\widetilde{B} - \hat{S} \\ (\widetilde{A}^\top \mathcal{X}\widetilde{B} - \hat{S})^\top & -\hat{R} + \widetilde{B}^\top \mathcal{X}\widetilde{B} \end{bmatrix} \begin{bmatrix} \xi_k \\ u_k \end{bmatrix} \leq 0 \tag{5.21}$$

for all $k$ and all trajectories $(u, \xi)$ of the extended system in (5.15) (with $\xi_0 \in \mathcal{X}_\xi$). Hence, there exists a lower bounded quadratic storage function $V(\xi_k) = \xi_k^\top \mathcal{X}\xi_k$ for which the extended system in (5.15) satisfies the dissipation inequality in (2.4), which directly implies that

$$\sum_{k=0}^{r} s(u_k, y_k) \geq V(\xi_r) - V(\xi_0) \geq 0$$

holds for all trajectories $\{u_k, y_k\}_{k=0}^\infty$ of the extended system with initial condition $\xi_0 = 0$. Since System (2.1) has the same input-output behavior as the extended system in (5.15), the condition in Theorem 2.2 is also fulfilled for System (2.1). Theorem 2.2 hence implies that System (2.1) is $(Q, S, R)$-dissipative.

(ii) We prove this direction via contraposition. If System (2.1) is $(Q, S, R)$-dissipative, then, according to Theorem 2.1, there exists a quadratic storage function $V(x_k) = x_k^\top \mathcal{X}' x_k$ such that

$$x_{k+1}^\top \mathcal{X}' x_{k+1} - x_k^\top \mathcal{X}' x_k \leq s(u_k, y_k)$$

holds for all $k$ and all $(u, x, y)$ satisfying (2.1). From the proof of Lemma 5.3, we know that there exists a transformation matrix $T$ such that $x_k = T\xi_k$ holds for all reachable states $\xi_k$ and all $k$. Hence, the matrix $\mathcal{X} = T^\top \mathcal{X}' T \succeq 0$ satisfies (5.21) for all $k$ and all $(u, \xi)$ of the extended system in (5.15). Using Willems' fundamental lemma [108], this implies that (5.20) holds and thus there exists a $\mathcal{X} \succeq 0$ such that (5.19) holds.

■

Theorem 5.3 provides an equivalent formulation of dissipativity based on input-output data. The result itself and its proof are conceptually similar to the case of state measurements in Theorem 5.1. A key challenge is that, in contrast to the matrix $\begin{bmatrix} X \\ U \end{bmatrix}$, the matrix $\begin{bmatrix} \Xi \\ U_\Xi \end{bmatrix}$ does usually not have full row rank, even if the input is persistently exciting, since the extended system in (5.15) is usually not controllable. However, Willems' fundamental lemma implies that, assuming the input to be persistently exciting of order $l+n$, the matrix $\Xi$ spans the space of all state trajectories of the extended system in (5.15). Under the stronger assumption of persistence of excitation of order $l+n+1$, which we assume in Theorem 5.3, it even holds that $\begin{bmatrix} \Xi \\ U_\Xi \end{bmatrix}$ spans the space of all input-state trajectories of (5.15). Using this fact, it is then straightforward to derive (5.19), which provides an equivalent data-driven characterization of dissipativity.

## 5.4 Dissipativity from noisy input-output trajectories

In this section, we extend the results of Section 5.3 to the case of noisy input-output data. From an input-output viewpoint, we consider System (2.1) in the difference operator form

$$\begin{aligned} y_k = &- a_l y_{k-1} - \cdots - a_2 y_{k-l+1} - a_1 y_{k-l} \\ &+ d u_k + b_l u_{k-1} + \cdots + b_2 u_{k-l+1} + b_1 u_{k-l}, \end{aligned} \tag{5.22}$$

with $a_i \in \mathbb{R}^{p\times p}$, $b_i \in \mathbb{R}^{p\times m}$, $i = 1,\ldots,l$, and $l$ being an upper bound on the lag $l \geq \underline{l}$ (cf. Lemma 5.3).

Instead of having exact measurements of the output, we assume that the input-output behavior is corrupted by process noise of the form

$$\begin{aligned} y_k = &- a_l y_{k-1} - \cdots - a_2 y_{k-l+1} - a_1 y_{k-l} \\ &+ d u_k + b_l u_{k-1} + \cdots + b_2 u_{k-l+1} + b_1 u_{k-l} + b_v v_k, \end{aligned} \tag{5.23}$$

where $v_k \in \mathbb{R}^{m_v}$ denotes the noise and, as before, $b_v \in \mathbb{R}^{p\times m_v}$ is a known matrix with full column rank, which can be used to include prior knowledge on the influence of the noise (cf. Remark 5.3). The noisy input-output behavior in (5.23) can also be represented in state-space via

$$\begin{bmatrix} u_{k-l+1} \\ \vdots \\ u_{k-1} \\ u_k \\ y_{k-l+1} \\ \vdots \\ y_{k-1} \\ y_k \end{bmatrix} = \begin{bmatrix} 0 & I & \dots & 0 & 0 & 0 & \dots & 0 \\ \vdots & \ddots & \ddots & \ddots & \vdots & & \ddots & \vdots \\ 0 & 0 & \dots & I & 0 & 0 & \dots & 0 \\ 0 & 0 & \dots & 0 & 0 & 0 & \dots & 0 \\ 0 & 0 & \dots & 0 & 0 & I & \dots & 0 \\ \vdots & \ddots & \ddots & \vdots & \vdots & & \ddots & \vdots \\ 0 & 0 & \dots & 0 & 0 & 0 & \dots & I \\ b_1 & b_2 & \dots & b_l & -a_1 & -a_2 & \dots & -a_l \end{bmatrix} \begin{bmatrix} u_{k-l} \\ u_{k-l+1} \\ \vdots \\ u_{k-1} \\ y_{k-l} \\ y_{k-l+1} \\ \vdots \\ y_{k-1} \end{bmatrix} + \begin{bmatrix} 0 \\ \vdots \\ 0 \\ I \\ 0 \\ \vdots \\ 0 \\ d \end{bmatrix} u_k + \begin{bmatrix} 0 \\ \vdots \\ 0 \\ 0 \\ 0 \\ \vdots \\ 0 \\ b_v \end{bmatrix} v_k. \tag{5.24}$$

Since only the last block row in (5.24) is uncertain, we introduce the notation

$$\begin{aligned} \xi_{k+1} &= \begin{bmatrix} \tilde{A}_1 \\ \tilde{A}_2 \end{bmatrix} \xi_k + \begin{bmatrix} \tilde{B}_1 \\ \tilde{D} \end{bmatrix} u_k + \begin{bmatrix} 0 \\ b_v \end{bmatrix} v_k, \\ y_k &= \tilde{A}_2 \xi_k + \tilde{D} u_k + b_v v_k, \end{aligned} \tag{5.25}$$

where $\tilde{A}_1 \in \mathbb{R}^{((p+m)l-p)\times(p+m)l}$ and $\tilde{B}_1 \in \mathbb{R}^{((p+m)l-p)\times m}$ are known (cf. (5.24)), and $\tilde{A}_2 \in \mathbb{R}^{p\times(p+m)l}$ and $\tilde{D} \in \mathbb{R}^{p\times m}$ are unknown.

With the input-output viewpoint in (5.22) and the extended state system representation in (5.24), we can follow a similar approach as in Section 5.2 to derive dissipativity conditions based on noisy input-output data. Similar as in the input-state case, we denote the actual noise sequence which yields the available input-output trajectory $\{u_k, y_k\}_{k=0}^{N-1}$ by $\{\hat{v}_k\}_{k=0}^{N-1}$. While the exact noise instance is unknown, we assume that we have information on a bound on the stacked matrix

$$\hat{V} = \begin{bmatrix} \hat{v}_l & \hat{v}_{l+1} & \dots & \hat{v}_{N-1} \end{bmatrix} \tag{5.26}$$

as specified in the following assumption.

**Assumption 5.2.** *The matrix $\hat{V}$ in (5.26) is an element of the set*

$$\mathcal{V} = \{V \in \mathbb{R}^{p\times(N-l)} | \begin{bmatrix} V^\top \\ I \end{bmatrix}^\top \begin{bmatrix} Q_v & S_v \\ S_v^\top & R_v \end{bmatrix} \begin{bmatrix} V^\top \\ I \end{bmatrix} \succeq 0\},$$

*where $Q_v \in \mathbb{R}^{(N-l)\times(N-l)}$, $S_v \in \mathbb{R}^{(N-l)\times m_v}$, and $R_v \in \mathbb{R}^{m_v\times m_v}$, with $Q_v \prec 0$.*

Due to the presence of noise, there generally exist multiple matrix pairs $(\widetilde{A}_2, \widetilde{D})$ which are consistent with the data for some noise sequence $V \in \mathcal{V}$. We denote the set of all such matrix pairs consistent with the input-output data and the noise bound by

$$\Sigma_{U,Y} = \{(\widetilde{A}_{2,\mathrm{d}}, \widetilde{D}_{\mathrm{d}}) | Y_{\Xi} = \widetilde{A}_{2,\mathrm{d}}\Xi + \widetilde{D}_{\mathrm{d}}U_{\Xi} + b_v V, V \in \mathcal{V}\}.$$

Along the lines of Section 5.2, this leads to an equivalent formulation for the set $\Sigma_{U,Y}$.

**Lemma 5.4.** *It holds that*

$$\Sigma_{U,Y} = \{(\widetilde{A}_{2,\mathrm{d}}, \widetilde{D}_{\mathrm{d}}) | \begin{bmatrix} \widetilde{A}_{2,\mathrm{d}}^\top \\ \widetilde{D}_{\mathrm{d}}^\top \\ I \end{bmatrix}^\top \begin{bmatrix} \bar{Q}_v & \bar{S}_v \\ \bar{S}_v^\top & \bar{R}_v \end{bmatrix} \begin{bmatrix} \widetilde{A}_{2,\mathrm{d}}^\top \\ \widetilde{D}_{\mathrm{d}}^\top \\ I \end{bmatrix} \succeq 0\} \tag{5.27}$$

*with*

$$\bar{Q}_v = \begin{bmatrix} \Xi \\ U_{\Xi} \end{bmatrix} Q_v \begin{bmatrix} \Xi \\ U_{\Xi} \end{bmatrix}^\top, \quad \bar{S}_v = -\begin{bmatrix} \Xi \\ U_{\Xi} \end{bmatrix} \left(Q_v Y_{\Xi}^\top + S_v b_v^\top\right),$$
$$\bar{R}_v = Y_{\Xi} Q_v Y_{\Xi}^\top + Y_{\Xi} S_v b_v^\top + b_v S_v^\top Y_{\Xi}^\top + b_v R_v b_v^\top.$$

*Proof.* With

$$\begin{bmatrix} V^\top \\ I \end{bmatrix} = \begin{bmatrix} -\Xi^\top & -U_{\Xi}^\top & Y_{\Xi}^\top \\ 0 & 0 & I \end{bmatrix} \begin{bmatrix} \widetilde{A}_{2,\mathrm{d}}^\top \\ \widetilde{D}_{\mathrm{d}}^\top \\ I \end{bmatrix},$$

this lemma can be proven similar to the proof of [102, Lemma 4 and Remark 2]. ∎

With the quadratic bound on the unknown matrices $(\widetilde{A}_{2,\mathrm{d}}, \widetilde{D}_{\mathrm{d}})$ from an input-output trajectory, we can again reformulate the uncertain system as an LFT of a nominal system

$$\begin{bmatrix} \xi_{k+1} \\ \tilde{z}_k \end{bmatrix} = \left[\begin{array}{ccc} \widetilde{A}_1 & \widetilde{B}_1 & 0 \\ 0 & 0 & I \\ \hline I & 0 & 0 \\ 0 & I & 0 \end{array}\right] \begin{bmatrix} \xi_k \\ u_k \\ \tilde{v}_k \end{bmatrix}, \quad \tilde{v}_k = \begin{bmatrix} \widetilde{A}_{2,\mathrm{d}} & \widetilde{D}_{\mathrm{d}} \end{bmatrix} \tilde{z}_k, \tag{5.28}$$

with $(\widetilde{A}_{2,\mathrm{d}}, \widetilde{D}_{\mathrm{d}}) \in \Sigma_{U,Y}$.

The reformulation in (5.28) allows us to apply robust analysis results, similar as in Section 5.2, to guarantee dissipativity properties from noisy input-output trajectories. For this, we assume again that the inverse in (5.13) exists.

**Theorem 5.4.** *Let $\tilde{R} \preceq 0$. If there exist a matrix $\mathcal{X} = \mathcal{X}^\top \succ 0$ and a scalar $\tau > 0$ such that*

$$\begin{bmatrix} \star & \star & \star \\ \star & \star & \star \\ \hline \star & \star & \star \\ \star & \star & \star \\ \hline \star & \star & \star \\ \star & \star & \star \end{bmatrix}^\top \left[\begin{array}{cc|cc|cc} -\mathcal{X} & 0 & 0 & 0 & 0 & 0 \\ 0 & \mathcal{X} & 0 & 0 & 0 & 0 \\ \hline 0 & 0 & -\tilde{R} & -\tilde{S}^\top & 0 & 0 \\ 0 & 0 & -\tilde{S} & -\tilde{Q} & 0 & 0 \\ \hline 0 & 0 & 0 & 0 & -\tau\bar{Q}_v & -\tau\bar{S}_v \\ 0 & 0 & 0 & 0 & -\tau\bar{S}_v^\top & -\tau\bar{R}_v \end{array}\right] \begin{bmatrix} \begin{bmatrix} \widetilde{A}_1^\top & 0 \end{bmatrix} & 0 & \begin{bmatrix} I & 0 \end{bmatrix} \\ -I & 0 & 0 \\ \begin{bmatrix} \widetilde{B}_1^\top & 0 \end{bmatrix} & 0 & \begin{bmatrix} 0 & I \end{bmatrix} \\ 0 & -I & 0 \\ 0 & 0 & I \\ \begin{bmatrix} 0 & I \end{bmatrix} & I & 0 \end{bmatrix} \succ 0 \tag{5.29}$$

*holds, then System (5.22) is $(Q, S, R)$-dissipative for all matrices consistent with the data $(\widetilde{A}_{2,\mathrm{d}}, \widetilde{D}_\mathrm{d}) \in \Sigma_{U,Y}$.*

*Proof.* With the result from Lemma 5.4, the proof follows along the arguments of the proof of Theorem 5.2. ∎

**Remark 5.7.** Note that from the proof of Lemma 5.3 (and the extended system as described in (5.18)), one can see that the matrices $a_i$, $b_i$, $i = 1, \ldots, l$ (5.22) are uniquely defined if the left inverse of $\mathcal{O}_l$ is unique, and hence, if $l = \underline{l}$ and $p\underline{l} = n$. Empirical evaluations showed that in these cases, the presented approach based on the condition in (5.29) worked very well in numerical examples, while for overapproximations of the lag no reasonable upper bounds on the respective dissipativity properties could be found. Improving the approach in these cases is part of future work.

**Example 5.2.** We illustrate the introduced approach with a numerical example. We consider a randomly generated system of order $n = 4$ with two inputs and two outputs $m = p = 2$. The system has an $\mathcal{L}_2$-gain of $\gamma_{\mathrm{tr}} = 3.73$. We assume knowledge of the lag $\underline{l} = 2$, and we simulate the trajectory with $u_k$, $k = 0, \ldots, N-1$ uniformly sampled in $[-1, 1]$, for different lengths $N$. We sample the noise $\hat{v}_k$ uniformly from the ball $\|\hat{v}_k\| \leq \bar{v}$ for all $k = 0, \ldots, N-1$ with $\bar{v} = 0.01$, which implies a bound on the measurement noise of $\hat{V}^\top \hat{V} \preceq \bar{v}^2(N - \underline{l})I$ (cf. Assumption 5.2). We then apply the result of Theorem 5.4, choose $\tilde{R} = \frac{1}{\gamma^2} I$, $\tilde{S} = 0$, $\tilde{Q} = -I$, and solve an SDP minimizing $-\frac{1}{\gamma^2}$ such that (5.29) holds. The resulting upper bound on the $\mathcal{L}_2$-gain, which is guaranteed for all systems that are consistent with the data, is depicted in Figure 5.3 for different data lengths. Next, we take the same example and increase the noise level $\bar{v}$ from 0.001 to 0.022 for $N = 50$ data points each and again apply the result from Theorem 5.4. The resulting upper bounds on the $\mathcal{L}_2$-gain are depicted in Figure 5.4.

The results in Figure 5.3 and Figure 5.4 are well aligned with the theoretical guarantees. The computed $\hat{\gamma}$ is indeed always an upper bound on the true $\mathcal{L}_2$-gain $\gamma_{\text{tr}}$ whenever such a bound can be found. Furthermore, similar qualitative behavior as in Example 5.1 can be observed: Increasing noise bounds lead to more conservatism, while more data points tend to improve the results in this example.

## 5.5 Experimental application example

In the following, we apply the presented results to an experimental setup to show the potential and applicability of the introduced ideas in real-world applications. More specifically, we apply the result from Section 5.2 to determine bounds on the $\mathcal{L}_2$-gain as well as passivity properties of a two-tank system locally around a steady state, and we compare the results to system identification approaches.

The experimental setup of the two-tank water system can be seen in Figure 5.5. It consists of two identical water tanks. The first water tank is fed by a water pump and the second water tank is fed by an outlet of the first water tank and has a water outlet itself. The considered input is the voltage $u_v$ which directly influences the throughput of the pump $v$. The heights of the two tanks $h_1$ and $h_2$ are considered the outputs, which can be measured.

By first principles, the two-tank can be modeled by

$$\begin{aligned}
\dot{h}_1 &= -\frac{O_1}{A_1}\mu_1\sqrt{2gh_1} + \frac{1}{A_1}k_{\text{p}}u_v, \\
\dot{h}_2 &= -\frac{O_2}{A_2}\mu_2\sqrt{2gh_2} + \frac{O_1}{A_2}\mu_1\sqrt{2gh_1},
\end{aligned}$$

where $A_i$, $O_i$ are the cross section and the outlet of tank $i = 1, 2$, respectively, $k_{\text{p}}$ is a constant of the pump including the tube and outlet, and $\mu_i$ captures approximately the hydro-dynamic resistance of the outlet of tank $i = 1, 2$.

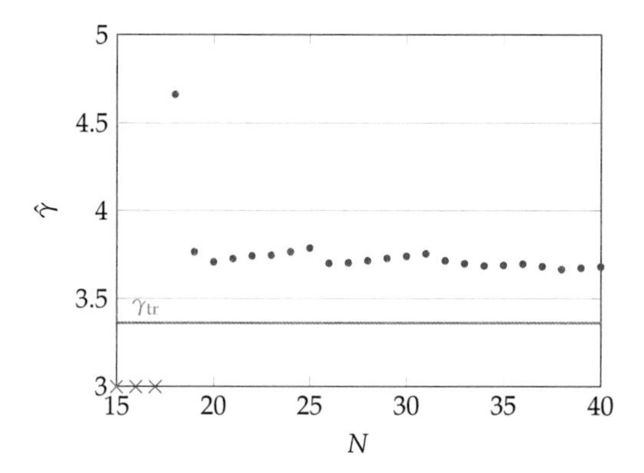


**Figure 5.3.** Guaranteed upper bounds on the $\mathcal{L}_2$-gain of the system in Example 5.2 from noisy input-output trajectories of different lengths $N$ (•). The red crosses ($\times$) indicate that no upper bound could be found.

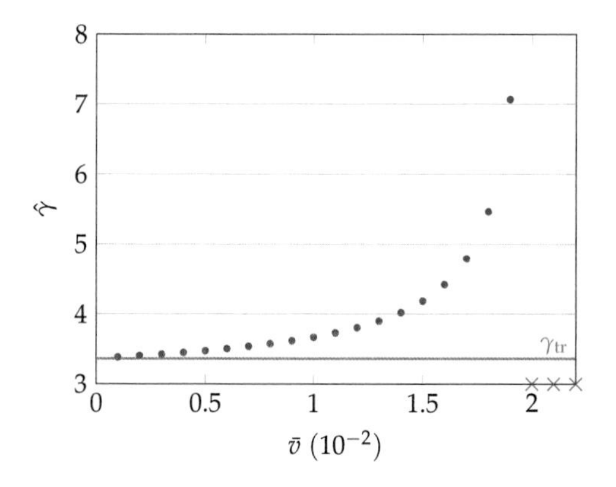


**Figure 5.4.** Guaranteed upper bounds on the $\mathcal{L}_2$-gain of the system in Example 5.2 from noisy input-output trajectories for increasing noise levels $\bar{v}$ and $N = 50$ data points (•). The red crosses ($\times$) indicate that no upper bound could be found.

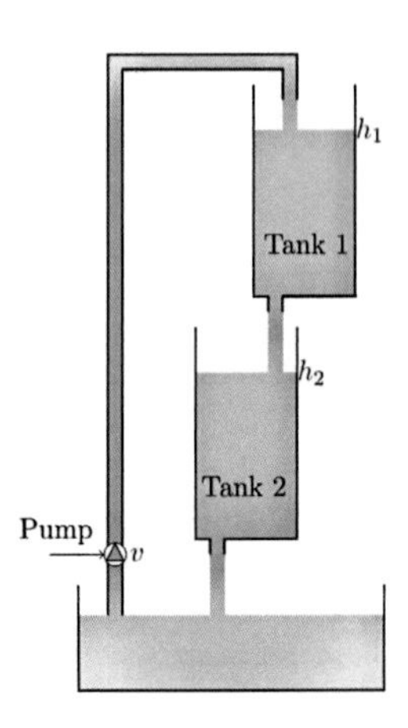


**Figure 5.5.** Experimental setup and schematic of the two-tank system.

Linearizing the system around a stationary point $(u_v^0, h_1^0, h_2^0)$ yields

$$\dot{x} = \begin{bmatrix} -\frac{\mu_1 O_1 \sqrt{2g}}{2A_1\sqrt{h_1^0}} & 0 \\ \frac{\mu_1 O_1 \sqrt{2g}}{2A_2\sqrt{h_1^0}} & -\frac{\mu_2 O_2 \sqrt{2g}}{2A_2\sqrt{h_2^0}} \end{bmatrix} x + \begin{bmatrix} \frac{k_p}{A_1} \\ 0 \end{bmatrix} u, \quad y = \begin{bmatrix} 1 & 0 \\ 0 & 1 \end{bmatrix} x, \tag{5.30}$$

with $u = u_v - u_v^0$, $x = \begin{bmatrix} h_1 - h_1^0 \\ h_2 - h_2^0 \end{bmatrix}$.

We are now interested in local information about the system around a setpoint. To determine a setpoint, we apply a simple controller in a first experiment that stabilizes the system at $h_2^0$. From this first experiment, we approximately determine the steady state $(u_v^0, h_1^0, h_2^0) = (6.8, 13.8, 16.4)$. In a second experiment, we excite the system around this steady state. The resulting input signal $u_v$ can be seen in Figure 5.6. We measure the resulting two heights $h_1$ and $h_2$ over 18 seconds with a sampling time of $T_s = 0.4$ seconds. The measured heights $h_1$ and $h_2$ are also shown in Figure 5.6.

As $g$ is a given physical constant and the parameters $A_i$ and $O_i$ are geometrical specifications of the experimental setup, we estimate the remaining parameters $k_p$, $\mu_1$ and $\mu_2$ in (5.30) via a least-squares approach (MATLAB function *lsqcurvefit*). The

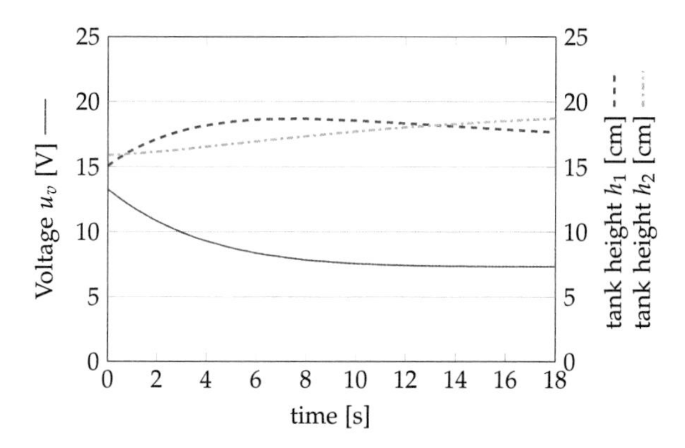


**Figure 5.6.** Input $u_v$ and the measured heights $h_1$ and $h_2$ of the two tank system over a horizon of 18 seconds and a sampling time of $T_s = 0.4$ seconds.

$\mathcal{L}_2$-gain of the resulting system is given by $\gamma_{\text{ParamEst}} = 6.85$. As an alternative, we compute $\begin{bmatrix} A & B \end{bmatrix}$ via

$$\begin{bmatrix} A & B \end{bmatrix} = X_+ \begin{bmatrix} X \\ U \end{bmatrix}^\dagger \tag{5.31}$$

where $^\dagger$ denotes the right inverse. The result can be interpreted as the result to the least-squares problem

$$\min_{\begin{bmatrix} A & B \end{bmatrix}} \|X_+ - \begin{bmatrix} A & B \end{bmatrix} \begin{bmatrix} X \\ U \end{bmatrix} \|_F,$$

where $\| \cdot \|_F$ indicates the Frobenius norm [75]. The resulting $\mathcal{L}_2$-gain of the identified system is given by $\gamma_{\text{LS}} = 5.20$, underestimating the $\mathcal{L}_2$-gain $\gamma_{\text{ParamEst}}$ by more than 20%.

Secondly, we are interested in the passivity properties of the system. More specifically, we want to determine the input feedforward passivity index $\nu$ for the input-output pair $u_v$ and $h_2$. The estimated input feedforward passivity indices obtained from first identifying a model and then applying model-based systems analysis tools are collected in Table 5.1, together with a summary of the $\mathcal{L}_2$-gain estimates.

**Table 5.1.** Estimates on the $\mathcal{L}_2$-gain and passivity from consecutive system identification and model-based analysis.

| | ParamEst | LS (Eq. (5.31)) |
|---|---|---|
| $\mathcal{L}_2$-gain $\hat{\gamma}$ | 6.85 | 5.20 |
| Input feedforward passivity index $\hat{\nu}$ | -0.515 | -0.588 |

Next, we apply the approach introduced in Section 5.2 to find guaranteed bounds on the $\mathcal{L}_2$-gain as well as the input feedforward passivity index from the available data. For this, we assume that the noise or disturbance in the data can be modeled by process noise that enters the linearized model in (5.30) as described in (5.4). We apply the results of Theorem 5.2 by assuming a bound on the process noise given by $\|w_k\| \leq \bar{w}$, which implies $\hat{W}\hat{W}^\top \preceq \bar{w}^2 NI$. The result is plotted in Figure 5.7 and Figure 5.8 for different assumed noise levels $\bar{w}$. In fact, we retrieve provably robust, reasonable upper bounds on the $\mathcal{L}_2$-gain of the system as well as lower bounds on the input feedforward passivity index.

We can see that for an assumed noise bound of $\bar{w} \leq 0.0077$, neither an $\mathcal{L}_2$-gain nor an input feedforward passivity index can be found. This indicates that there does not exist any LTI system which admits an $\mathcal{L}_2$-gain $\gamma \in \mathbb{R}_{\geq 0}$ or feedforward passivity index $\nu \in \mathbb{R}$, respectively, that is consistent with the data assuming such a low noise level. For an assumed noise level of $\bar{w} = 0.008$ the proposed approach yields an $\mathcal{L}_2$-gain of $\hat{\gamma} = 7.92$ and an input feedforward passivity index of $\hat{\nu} = -0.99$. When comparing these values to the baseline *ParamEst*, which uses physical insights together with parameter estimation, the results are more conservative. This is to be expected as the resulting upper (or lower) bound is the system property that is guaranteed for all systems consistent with the data and the assumed noise model. In particular, we note that the approaches based on system identification methods do not provide any theoretical guarantees on the actual system property satisfied by the two-tank system. The assumed noise level of course highly influences the result, as indicated also in Figure 5.7 and Figure 5.8. However, as too small noise levels lead to infeasibility, the data already implicitly reveal a reasonable interval for the noise bound. Naturally, for larger assumed noise bounds, the estimated bound on the system property becomes more conservative, since the set of systems that are consistent with the data and the noise bound increases.

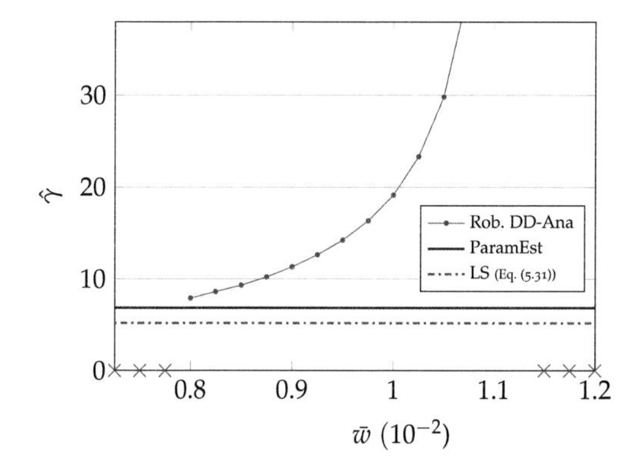


**Figure 5.7.** Upper bound on the $\mathcal{L}_2$-gain of the two-tank system from noisy input-state trajectories for different bounds on the process noise $\bar{w}$. The blue dots • are the upper bounds computed by the introduced robust data-driven analysis approach, the red crosses × indicate that no upper bound could be found, and the differently colored lines indicate the results of consecutive system identification and systems analysis as specified by the legend.

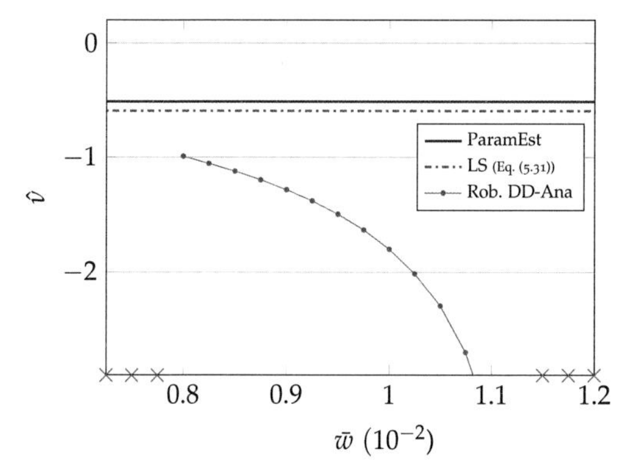


**Figure 5.8.** Lower bound on the input feedforward passivity index of the two-tank system from noisy input-state trajectories for different bounds on the process noise $\bar{w}$. The blue dots • are the lower bounds computed by the introduced robust data-driven analysis approach, the red crosses × indicate that no lower bound could be found, and the differently colored lines indicate the results of consecutive system identification and systems analysis as specified by the legend.

Altogether, the presented results show the potential and applicability of the introduced approach for real experimental measurements. For a given noise level, the results provide provably robust bounds on dissipativity properties over the infinite horizon for all systems consistent with the data. By varying the assumed noise bound, the introduced framework allows for an intuitive trade-off between accuracy of the estimated system property and robustness (i.e., the size of the set of systems for which the property is guaranteed).

## 5.6 Summary

In this chapter, we introduced simple verification methods of dissipativity properties with guarantees from (noisy) data of finite length via computationally attractive LMI-based conditions. We started by introducing a dissipativity characterization on the basis of one input-state trajectory leading to a simple semidefiniteness condition of a single data-dependent matrix. We then extended this result via robust analysis tools to guarantee dissipativity properties from noisy data given deterministic bounds on the process noise. Furthermore, we introduced approaches to verify dissipativity from one input-output trajectory both with and without noise by utilizing an artificial extended state based on the given input-output data. The applicability of the presented methods were shown both in small numerical examples as well as in an experimental application example.

In contrast to the methods presented in Chapter 3 and Chapter 4, the difference viewpoint in this chapter, exploiting Theorem 2.1, allows to directly infer dissipativity properties over the infinite horizon from data of finite length. Furthermore, while in Chapter 3 and Chapter 4 no or at most qualitative guarantees in the case of noisy data can be given, the robust viewpoint in this chapter, utilizing tools and results from robust analysis and control, provides rigorous guarantees in the case of unknown but bounded process noise.

It is part of future research to decrease the conservatism by considering parameter-dependent storage functions as well as to extend the introduced approach in the case of noisy input-output trajectories to find a tight description of the system properties given conservative upper bounds on the lag. While the ideas presented in this chapter are readily extended to verify dissipativity properties of polynomial systems [55] and of interconnected systems [92], it might be interesting for future work to investigate how the presented ideas can be extended to further classes of nonlinear systems.

# Chapter 6

# Conclusions

In this chapter, we recall and summarize the main results of this thesis, discuss the respective advantages of the three complementary approaches to determine input-output system properties from data as proposed in the Chapters 3, 4, and 5, and point to a few further questions for future research. This chapter concludes the thesis.

## 6.1 Summary

This thesis addressed the problem of determining input-output properties of LTI systems directly from data while no mathematical model of the system is known. To this end, we presented three complementary data-driven analysis approaches, each providing a simple application scheme together with a rigorous theoretical foundation. The first framework in Chapter 3 encompasses active sampling schemes, where iterative experiments on the unknown systems reveal the system property of interest. In Chapter 4, we introduced an offline method requiring only one input-output trajectory by applying Willems' fundamental lemma to data-driven analysis. Lastly, in Chapter 5, we switched to a robust viewpoint to develop an approach to data-driven systems analysis that provides rigorous guarantees in the case that the data is corrupted by bounded process noise.

**Iterative methods.** Chapter 3 was devoted to iterative sampling schemes, where we extended a state-of-the-art approach to determine the $\mathcal{L}_2$-gain of an LTI system to a more general framework, which can be summarized by (i) writing the input-output system property in terms of an optimization problem, and (ii) iteratively updating the input along the gradient of the optimization problem obtained from data. The resulting sampling schemes to determine the $\mathcal{L}_2$-gain, passivity properties, and conic relations are thus based on gradient dynamical systems and saddle point flows, for which the convergence results are stated in Theorem 3.1, Theorem 3.2, Theorem 3.3, and Theorem 3.4, respectively. While Proposition 3.1, Proposition 3.3, and Proposition 3.5 show that the respective gradients can be obtained from input-output trajectories, the two adapted sampling schemes analyzed in Proposition 3.4 and Proposition 3.6 improve data efficiency in the case of passivity properties and conic relations. Proposition 3.7 and Proposition 3.8 then address how the general framework for iteratively determining certain input-output properties from Section 3.1 can be extended to continuous-time LTI systems. Moreover, Section 3.2.2 generalizes the findings of Section 3.1 to MIMO systems, and Proposition 3.9 together with Lemma 3.4 provide probabilistic guarantees in the case of Gaussian measurement noise.

**Offline methods.** In Chapter 4, we introduced an offline method to determine input-output properties from one input-output trajectory utilizing Willems' fundamental lemma. The key concept of the chapter is reflected by Theorem 4.2, which provides an easily verifiable, necessary and sufficient semidefiniteness condition to certify IQCs requiring only an upper bound on the lag of the system as well as a persistently exciting input trajectory. Moreover, Proposition 4.1 allows to extend this result to finding not only the system property of interest but also the input which attains the bound on this system property. In the case of measurement noise, a heuristic relaxation for the verification of IQCs is provided in Section 4.1.1, which resulted in promising results in numerical examples. Theorem 4.3 in Section 4.2 extends the findings of Section 4.1 to determining optimal input-output properties via an SDP that provides the tightest description of the unknown system within a class of IQCs. Examples in this regard include the smallest cone containing the input-output behavior as well as finding a lower-order surrogate model. Thereafter, Section 4.3 is devoted to the question of deriving an upper bound on the difference between system properties over the infinite and the finite horizon, resulting in bounds captured in Theorem 4.4 and Proposition 4.2 for SISO systems and certain classes of MIMO systems, respectively.

**Robust methods.** Since no quantitative bounds on input-output properties in the case of unknown but deterministically bounded noise were provided in Chapter 3 and Chapter 4, Chapter 5 was devoted to finding rigorous guarantees on dissipativity properties in the case of noise-corrupted data via a robust viewpoint. We started by presenting a necessary and sufficient condition for dissipativity from one noise-free input-state trajectory in Theorem 5.1 based on a simple SDP, which we then extended to the case of input-state trajectories that are corrupted by unknown but bounded process noise. This resulted in the main result of this chapter, stated in Theorem 5.2, which introduces a nonconservative and computationally attractive condition to guarantee dissipativity properties from one noise-corrupted input-state trajectory of finite length. Thereafter, we extend the results from Section 5.1 and Section 5.2 to the case of input-output trajectories. Firstly, we utilized an extended state based on input-output data together with Willems' fundamental lemma to provide a necessary and sufficient condition for dissipativity from one noise-free input-output trajectory of finite length, which is summarized in Theorem 5.3. Along the ideas of Section 5.2, Theorem 5.3 is then extended to the case of input-output data corrupted by bounded process noise, which yields a simple SDP to certify dissipativity over the infinite horizon from one noise-corrupted and finite-length input-output trajectory, as formalized in Theorem 5.4.

## 6.2 Discussion

The three approaches to determining input-output system properties of LTI systems from data presented in this thesis are all designed to be simple in their application and based on a rigorous mathematical analysis. However, each of the three approaches comes with its own advantages, requirements and limitations, which we will shortly discuss in the following.

The requirements on the data are minimal for all three approaches. For both offline methods, introduced in Chapter 4 and Chapter 5, the only requirement on the data is some form of persistency of excitation condition on the input trajectory of the available data, and convergence of the iterative sampling schemes, as introduced in Chapter 3, can be shown for almost all initial input trajectories in the case of the $\mathcal{L}_2$-gain and passivity properties. Note that, while the sampling scheme for determining conic relations in Section 3.1.3 only comes with local convergence guarantees, Proposition 4.1 can be used to initialize the iterative scheme close to the optimizer. While for the iterative schemes introduced in Chapter 3 no additional prior knowledge is required, both offline approaches require an upper bound on the lag of the system, which we however do not consider a restrictive assumption.

The most significant drawback of the iterative framework from Chapter 3 is that iterative experiments are required, which can be time consuming for some applications. The additionally required computations, on the other hand, are negligible. Both offline methods, in contrast, require only one available trajectory, and verifying or determining an optimal system property yields a semidefiniteness condition or an SDP, respectively. Since there exist powerful and efficient algorithms to verify a semidefiniteness condition and to solve an SDP, the resulting computational expenses can still be considered low and hence computationally attractive.

Another difference between the three methods is that the two methods from Chapter 3 and Chapter 4 are based on the input-output condition of dissipativity (and IQCs) as provided in Theorem 2.2 (and Definition 2.6). As any application can only provide trajectories of finite length, we then consider $L$-dissipativity (and $L$-IQCs) as defined in Definition 2.5 (and Definition 2.7). While the system property over the infinite horizon is well approximated for large horizons $L$, and a theoretical bound on the difference is derived in Section 4.3, no tight bounds on the infinite horizon properties can be provided. The robust approach in Chapter 5, in contrast, utilizes the condition for dissipativity as stated in Theorem 2.1, and hence directly yields sharp results for the infinite horizon. Moreover, the robust method provides guaranteed bounds on dissipativity properties from measured trajectories that are corrupted by process noise, which offers a significant advantage compared to other approaches for data-driven systems analysis. However, the worst-case viewpoint of robust approaches generally only allows to handle rather small noise bounds, while the methods in Chapter 3 and Chapter 4 resulted in very promising results even for high noise levels and high dimensional systems. An open problem in this regard lies in providing guarantees in the case of noise-corrupted data for the offline method in Chapter 4, while the iterative schemes in Chapter 3 at least allow for qualitative guarantees and bounds as presented in Section 3.2.3. Note that initializing the presented iterative schemes with the results from Chapter 4 (Proposition 4.1) may provide a good joint approach in case of high measurement noise levels with improved data efficiency.

Finally, compared to the alternative of first identifying the system from data and applying model-based analysis methods to the resulting parametric mathematical model, all three data-driven approaches can be considered non-parametric. To be more precise, most standard system identification methods require knowledge of the order of the system, which can influence the results significantly (cf. Figure 4.4).

For the data-driven approaches in Chapter 3 and Chapter 4, no or only an upper bound on the lag or the order of the system needs to be known, which can simply be chosen almost arbitrarily conservatively. Furthermore, the examples in Section 3.3.3 and Section 4.4.1 exemplarily showed that the iterative approach from Chapter 3 and the offline method from Chapter 4 provided good estimates in the case of high-dimensional systems with high noise levels, while standard system identification methods with consecutive analysis produced quite variable results (cf. Figure 4.4, [AK8]). Lastly, developing system identification methods with guarantees from noisy data of finite length is part of ongoing research with some interesting initial results under certain assumptions (cf. Section 1.1). In contrast, the results in Chapter 5 provide computationally attractive approaches to guarantee quantitative bounds on dissipativity properties from noise-corrupted and finite-length data given deterministic noise bounds.

Altogether, all three presented approaches from Chapter 3, Chapter 4, and Chapter 5 are simple to apply, have little requirements on the data or with regard to prior knowledge, and demonstrated their applicability and potential in multiple (numerical) examples, also in comparison to standard system identification tools. Since the underlying concepts and consequent methods presented in the respective chapters are based on different control-theoretic principles and results, all three approaches demonstrate their distinct advantages and limitations, resulting in their individual application cases.

Systems analysis not only provides insights to the system and allows to do controller design with guarantees, but it can also validate a given controller or its closed-loop performance. Therefore, this thesis addressed the issue of determining input-output properties directly from data by presenting three distinct approaches based on rigorous mathematical analysis, and hence, contributed to a sound mathematical framework for data-driven system analysis and control theory.

## 6.3 Outlook

For each of the approaches in Chapter 3, Chapter 4, and Chapter 5, there are distinct open problems and ideas to improve the individual methods, as discussed in more detail in the individual chapters. The overarching open question with respect to all introduced methods is their extension to nonlinear systems.

With regard to iterative schemes, [38] illustrated by means of a numerical example that the power iteration method, developed for LTI systems, also yields a good estimate on the $\mathcal{L}_2$-gain for certain nonlinear systems. To analyze this observation from a theoretical viewpoint, considering not only the $\mathcal{L}_2$-gain but also other input-output properties, would be an interesting topic for future research. A more specific idea to generalize the underlying scheme of Chapter 3 to a class of nonlinear systems is to consider Hamiltonian systems and exploit their self-adjoint property. To be more precise, with results from [28], gradient information of the derived optimization problems corresponding to certain system properties can again be obtained from input-output trajectories with a very similar time-reversal trick as in the LTI case. For more general nonlinear systems, gradient information can also be obtained on the basis of extremum seeking schemes (see, e.g., [25]).

For the offline approach to infer IQCs as introduced in Chapter 4, an open problem is to find rigorous guarantees in the case of noisy data. Since this approach is based on Willems' fundamental lemma, it is additionally very interesting to investigate whether and how initial ideas on the extension of Willems' fundamental lemma to, e.g., Hammerstein or Wiener systems [9], linear parameter-varying (LPV) systems [AK25], or more general nonlinear systems via kernel based approaches [46] could be utilized to develop a simple verification scheme for IQCs from data for such nonlinear systems. When taking a step further away from Willems' fundamental lemma, another idea could be to utilize so-called scenario-based results [16, 29] to infer probabilistic guarantees over all possible input-output trajectories from only a finite set of offline available data trajectories.

Immediate open questions for future research regarding the robust approach presented in Chapter 5 include the extension to parameter-dependent storage functions as well as finding a tight description of the system properties in the case of noisy input-output trajectories given a conservative upper bound on the lag. As this robust approach to dissipativity from data is based on an LTI formulation of dissipativity, first extensions to nonlinear systems include substitution or lifting methods which allow to recover the LTI structure. An example in this direction can be found in [55], where dissipativity properties from noisy input-state trajectories of polynomial systems can be verified. Similar ideas could possibly be merged with nonlinear system descriptions as presented for Hammerstein and Wiener systems [9], second-order Volterra systems [83], and rational dynamics [93].

# Bibliography


[1] P.-A. Absil, R. Mahony, and R. Sepulchre. *Optimization algorithms on matrix manifolds*. Princeton University Press, 2008.

[2] M. Arcak, C. Meissen, and A. K. Packard. *Networks of dissipative systems*. Springer, 2016.

[3] K. J. Arrow, L. Hurwicz, and H. Uzawa. *Studies in linear and non-linear programming*. Stanford University Press, 1958.

[4] G. Baggio, D. S. Bassett, and F. Pasqualetti. "Data-driven control of complex networks." In: *Nature communications* 12.1 (2021), pp. 1–13.

[5] S. Batterson and J. Smillie. "Dynamics of Rayleigh quotient iteration." In: *SIAM J. Numer. Anal.* 26.3 (1989), pp. 624–636.

[6] S. Beheshti and M. A. Dahleh. "A new information-theoretic approach to signal denoising and best basis selection." In: *IEEE Trans. Signal Process.* 53.10 (2005), pp. 3613–3624.

[7] M. Benzi, G. H. Golub, and J. Liesen. "Numerical solution of saddle point problems." In: *Acta Numerica* 14 (2005), pp. 1–137.

[8] J. Berberich, J. Köhler, M. A. Müller, and F. Allgöwer. "Data-driven model predictive control with stability and robustness guarantees." In: *IEEE Trans. Automat. Control* 66.4 (2021), pp. 1702–1717.

[9] J. Berberich and F. Allgöwer. "A trajectory-based framework for data-driven system analysis and control." In: *Proc. European Control Conf.* 2020, pp. 1365–1370.

[10] J. Berberich, C. W. Scherer, and F. Allgöwer. "Combining prior knowledge and data for robust controller design." In: *arXiv preprint arXiv:2009.05253* (2020).

[11] A. Berndt, A. Alanwar, K. H. Johansson, and H. Sandberg. "Data-driven set-based estimation using matrix zonotopes with set containment guarantees." In: *arXiv preprint arXiv:2101.10784* (2021).

[12] R. Bhatia. *Matrix analysis*. Springer, 1997.

[13] A. Böttcher and S. M. Grudsky. *Toeplitz matrices, asymptotic linear algebra and functional analysis*. Springer, 2000.

[14] D. Bristow, M. Tharayil, and A. Alleyne. "A survey of iterative learning control." In: *IEEE Control Systems Magazine* 26.3 (2006), pp. 96–114.

[15] M. Cadic, J. W. Polderman, and I. M. Y. Mareels. "Set membership identification for adaptive control: Input design." In: *Proc. 42nd IEEE Conf. Decision and Control*. 2003, pp. 5011–5026.

[16] M. C. Campi, S. Garatti, and M. Prandini. "The scenario approach for systems and control design." In: *Annual Reviews in Control* 33.2 (2009), pp. 149–157.

[17] M. C. Campi, A. Lecchini, and S. M. Savaresi. "Virtual reference feedback tuning: A direct method for the design of feedback controllers." In: *Automatica* 38.8 (2002), pp. 1337–1346.

[18] J. Carrasco and P. Seiler. "Conditions for the equivalence between IQC and graph separation stability results." In: *Int. J. Control* 92.12 (2019), pp. 2899–2906.

[19] Y. Chahlaoui and P. V. Dooren. *A collection of benchmark examples for model reduction of linear time invariant dynamical systems*. SLICOT Working Note. 2002.

[20] J. Coulson, J. Lygeros, and F. Dörfler. "Data-enabled predictive control: In the shallows of the DeePC." In: *Proc. European Control Conf.* 2019, pp. 307–312.

[21] S. Dean, H. Mania, N. Matni, B. Recht, and S. Tu. "On the sample complexity of the linear quadratic regulator." In: *Found. Comput. Math.* 20.4 (2020), pp. 633–679.

[22] C. A. Desoer and M. Vidyasagar. *Feedback systems: Input-output properties*. SIAM, 1975.

[23] S. G. Dietz and C. W. Scherer. "Robust output feedback control against disturbance filter uncertainty described by dynamic integral quadratic constraints." In: *Int. J. Robust Nonlinear Control* 20.17 (2010), pp. 1903–1919.

[24] A. Falsone, F. Molinari, and M. Prandini. "Uncertain multi-agent MILPs: A data-driven decentralized solution with probabilistic feasibility guarantees." In: *Proc. 2nd Conf. Learning for Dynamics and Control*. 2020, pp. 1000–1009.

[25] J. Feiling, A. Zeller, and C. Ebenbauer. "Derivative-free optimization algorithms based on non-commutative maps." In: *IEEE Control Systems Letters* 2.4 (2018), pp. 743–748.

[26] R. Fletcher. "Semidefinite matrix constraints in optimization." In: *SIAM J. Control Optim.* 23.4 (1985), pp. 493–513.

[27] J. M. Fry, M. Farhood, and P. Seiler. "IQC-based robustness analysis of discrete-time linear time-varying systems." In: *Int. J. Robust Nonlinear Control* 27.16 (2017), pp. 3135–3157.

[28] K. Fujimoto and T. Sugie. "Iterative learning control of Hamiltonian systems: I/O based optimal control approach." In: *IEEE Trans. Automat. Control* 48.10 (2003), pp. 1756–1761.

[29] S. Garatti and M. C. Campi. "Risk and complexity in scenario optimization." In: *Mathematical Programming* (2019), pp. 1–37.

[30] T. T. Georgiou and M. C. Smith. "Optimal robustness in the gap metric." In: *IEEE Trans. Automat. Control* 35.6 (1990), pp. 673–686.

[31] G. H. Golub and C. F. van Loan. *Matrix computations*. Johns Hopkins University Press, 1996.

[32] U. Helmke and J. B. Moore. *Optimization and dynamical systems*. Springer, 1996.

[33] M. Herceg, M. Kvasnica, C. Jones, and M. Morari. "Multi-Parametric Toolbox 3.0." In: *Proc. European Control Conf.* 2013, pp. 502–510.

[34] M. R. Hestenes and W. Karush. "A method of gradients for the calculation of the characteristic roots and vectors of a real symmetric matrix." In: *J. Res. Natl. Bur. Stand.* 47.1 (1951), pp. 45–61.

[35] M. R. Hestenes and W. Karush. "Solutions of $Ax = \lambda Bx$." In: *J. Res. Natl. Bur. Stand.* 47.6 (1951), pp. 471–478.

[36] K. van Heusden, A. Karimi, and D. Bonvin. "Data-driven controller validation." In: *Proc. 15th IFAC Symp. System Identification*. 2009, pp. 1050–1055.

[37] D. J. Hill and P. J. Moylan. "Dissipative dynamical systems: Basic input-output and state properties." In: *J. Franklin Inst.* 309.5 (1980), pp. 327–357.

[38] H. Hjalmarsson. "From experiment design to closed-loop control." In: *Automatica* 41.3 (2005), pp. 393–438.

[39] H. Hjalmarsson. "Iterative feedback tuning - an overview." In: *Int. J. Adapt. Control Signal Processing* 16.5 (2002), pp. 373–395.

[40] Z.-S. Hou and Z. Wang. "From model-based control to data-driven control: Survey, classification and perspective." In: *Information Sciences* 235 (2013), pp. 3–35.

[41] B. Hu, M. J. Lacerda, and P. Seiler. "Robust analysis of uncertain discrete-time systems with dissipation inequalities and integral quadratic constraints." In: *Int. J. Robust Nonlinear Control* 27.11 (2017), pp. 1940–1962.

[42] B. Hu and P. Seiler. "Exponential decay rate conditions for uncertain linear systems using integral quadratic constraints." In: *IEEE Trans. Automat. Control* 61.11 (2016), pp. 3631–3636.

[43] K. Iijima, M. Tanemura, S.-i. Azuma, and Y. Chida. "Reduction in the amount of data for data-driven passivity estimation." In: *Proc. IEEE Conf. Control Technology and Applications*. 2020, pp. 134–139.

[44] E. de Klerk, F. Glineur, and A. B. Taylor. "On the worst-case complexity of the gradient method with exact line search for smooth strongly convex functions." In: *Optimization Letters* 11.7 (2017), pp. 1185–1199.

[45] R. L. Kosut. "Uncertainty model unfalsification for robust adaptive control." In: *Annual Reviews in Control* 25 (2001), pp. 65–76.

[46] Y. Lian and C. N. Jones. "Nonlinear data-enabled prediction and control." In: *Proc. 3rd Conf. Learning for Dynamics and Control*. 2021, pp. 523–534.

[47] L. Ljung. *System identification: Theory for the user*. Prentice-Hall, Inc., 1999.

[48] J. Löfberg. "YALMIP: A toolbox for modeling and optimization in MATLAB." In: *Proc. IEEE Int. Conf. Robotics and Automation*. 2004, pp. 284–289.

[49] R. Mahony and P.-A. Absil. "The continuous-time Rayleigh quotient flow on the sphere." In: *Linear Algebra Appl.* 368 (2003), pp. 343–357.

[50] A. Marco, P. Hennig, J. Bohg, S. Schaal, and S. Trimpe. "Automatic LQR tuning based on Gaussian process global optimization." In: *Proc. IEEE Int. Conf. Robotics and Automation*. 2016, pp. 270–277.

[51] I. M. Y. Mareels. "Sufficiency of excitation." In: *Systems & Control Letters* 5.3 (1984), pp. 159–163.

[52] I. Markovsky and P. Rapisarda. "Data-driven simulation and control." In: *Int. J. Control* 81.12 (2008), pp. 1946–1959.

[53] T. Martin and F. Allgöwer. "Iterative data-driven inference of nonlinearity measures via successive graph approximation." In: *Proc. 59th IEEE Conf. Decision and Control*. 2020, pp. 4760–4765.

[54] T. Martin and F. Allgöwer. "Nonlinearity measures for data-driven system analysis and control." In: *Proc. 58th IEEE Conf. Decision and Control*. 2019, pp. 3605–3610.

[55] T. Martin and F. Allgöwer. "Dissipativity verification with guarantees for polynomial systems from noisy input-state data." In: *IEEE Control Systems Letters* 5.4 (2020), pp. 1399–1404.

[56] N. Matni and S. Tu. "A tutorial on concentration bounds for system identification." In: *Proc. 58th IEEE Conf. Decision and Control*. 2019, pp. 3741–3749.

[57] T. Maupong, J. Mayo-Maldonado, and P. Rapisarda. "On Lyapunov functions and data-driven dissipativity." In: *Proc. 20th IFAC World Congress*. 2017, pp. 7783–7788.

[58] A. Megretski and A. Rantzer. "System analysis via integral quadratic constraints." In: *IEEE Trans. Automat. Control* 42.6 (1997), pp. 819–830.

[59] E. Mengi, E. A. Yildirim, and M. Kiliç. "Numerical optimization of eigenvalues of Hermitian matrix functions." In: *SIAM J. Matrix Anal. Appl.* 35.2 (2014), pp. 699–724.

[60] S. Michalowsky, C. Scherer, and C. Ebenbauer. "Robust and structure exploiting optimization algorithms: An integral quadratic constraint approach." In: *Int. J. Control* (2020), pp. 1–24.

[61] M. Milanese and A. Vicino. "Optimal estimation theory for dynamic systems with set membership uncertainty: An overview." In: *Automatica* 27.6 (1991), pp. 997–1009.

[62] J. M. Montenbruck and F. Allgöwer. "Some problems arising in controller design from big data via input-output methods." In: *Proc. 55th IEEE Conf. Decision and Control*. 2016, pp. 6525–6530.

[63] M. I. Müller and C. R. Rojas. "Gain estimation of linear dynamical systems using Thompson sampling." In: *Proc. 22nd Int. Conf. Artificial Intelligence and Statistics*. 2019, pp. 1535–1543.

[64] M. I. Müller, P. E. Valenzuela, A. Proutiere, and C. R. Rojas. "A stochastic multi-armed bandit approach to nonparametric $H_\infty$ estimation." In: *Proc. 56th IEEE Conf. Decision and Control*. 2017, pp. 4632–4637.

[65] M. I. Müller and C. R. Rojas. "Iterative $H_\infty$-norm estimation using cyclic-prefixed signals." In: *Proc. 59th IEEE Conf. Decision and Control*. 2020, pp. 2869–2874.

[66] M. Neumann-Brosig, A. Marco, D. Schwarzmann, and S. Trimpe. "Data-efficient autotuning with Bayesian optimization: An industrial control study." In: *IEEE Trans. Control Syst. Technol.* 28.3 (2020), pp. 730–740.

[67] M. Norrlöf and S. Gunnarsson. "Time and frequency domain convergence properties in iterative learning control." In: *Int. J. Control* 75.14 (2002), pp. 1114–1126.

[68] C. Novara and S. Formentin. "Data-driven inversion-based control of nonlinear systems with guaranteed closed-loop stability." In: *IEEE Trans. Automat. Control* 63.4 (2017), pp. 1147–1154.

[69] C. Novara and M. Milanese. "Control of MIMO nonlinear systems: A data-driven model inversion approach." In: *Automatica* 101 (2019), pp. 417–430.

[70] E. Oja. "Simplified neuron model as a principal component analyzer." In: *J. Math. Biol.* 15.3 (1982), pp. 267–273.

[71] M. C. de Oliveira and R. E. Skelton. "Stability tests for constrained linear systems." In: *Perspectives in Robust Control*. Springer, 2001, pp. 241–257.

[72] T. Oomen, R. van der Maas, C. R. Rojas, and H. Hjalmarsson. "Iterative data-driven $\mathcal{H}_\infty$ norm estimation of multivariable systems with application to robust active vibration isolation." In: *IEEE Trans. Control Syst. Technol.* 22.6 (2014), pp. 2247–2260.

[73] A. Ostrowski and H. Schneider. "Some theorems on the inertia of general matrices." In: *J. Math. Anal. Appl.* 4.1 (1962), pp. 72–84.

[74] S. Oymak and N. Ozay. "Non-asymptotic identification of LTI systems from a single trajectory." In: *Proc. American Control Conf.* 2019, pp. 5655–5661.

[75] C. D. Persis and P. Tesi. "Formulas for data-driven control: Stabilization, optimality and robustness." In: *IEEE Trans. Automat. Control* 65.3 (2020), pp. 909–924.

[76] B. T. Polyak. *Introduction to optimization.* Optimization Software, Inc., 1987.

[77] B. T. Polyak. "Iterative methods using Lagrange multipliers for solving extremal problems with constraints of the equation type." In: *USSR Comput. Math. & Math. Phys.* 10.5 (1970), pp. 42–52.

[78] L. D. Popov. "A modification of the Arrow-Hurwicz method for search of saddle points." In: *Mat. Zametki* 28.5 (1980), pp. 777–784.

[79] G. Rallo, S. Formentin, C. R. Rojas, T. Oomen, and S. M. Savaresi. "Data-driven $H_\infty$-norm estimation via expert advice." In: *Proc. 56th IEEE Conf. Decision and Control.* 2017, pp. 1560–1565.

[80] F. Rellich. *Perturbation theory of eigenvalue problems.* Gordon and Breach, 1969.

[81] A. Rogozhin. "The singular value behavior of the finite sections of block Toeplitz operators." In: *SIAM J. Matrix Anal. Appl.* 27.1 (2005), pp. 273–293.

[82] C. R. Rojas, T. Oomen, H. Hjalmarsson, and B. Wahlberg. "Analyzing iterations in identification with application to nonparametric $H_\infty$-norm estimation." In: *Automatica* 48.11 (2012), pp. 2776–2790.

[83] G. Rueda-Escobedo and J. Schiffer. "Data-driven internal model control of second-order discrete Volterra Systems." In: *Proc. 59th IEEE Conf. Decision and Control.* 2020, pp. 4572–4579.

[84] A. K. El-Sakkary. "The gap metric: Robustness of stabilization of feedback systems." In: *IEEE Trans. Automat. Control* 30.3 (1985), pp. 240–247.

[85] G. Scarciotti and A. Astolfi. "Data-driven model reduction by moment matching for linear and nonlinear systems." In: *Automatica* 79 (2017), pp. 340–351.

[86] C. W. Scherer. "LPV control and full block multipliers." In: *Automatica* 37.3 (2001), pp. 361–375.

[87] C. W. Scherer and S. Weiland. "Linear matrix inequalities in control." In: *Lecture Notes, Dutch Institute for Systems and Control, Delft, The Netherlands* 3.2 (2000).

[88] T. Schweickhardt and F. Allgöwer. "Linear control of nonlinear systems based on nonlinearity measures." In: *J. Proc. Contr.* 17.3 (2007), pp. 273–284.

[89] M. Sharf. "On the sample complexity of data-driven inference of the $\mathcal{L}_2$-gain." In: *IEEE Control Systems Letters* 4.4 (2020), pp. 904–909.

[90] A. H. Siddiqi. *Functional analysis and applications.* Springer, 2018.

[91] M. Simchowitz, H. Mania, S. Tu, M. I. Jordan, and B. Recht. "Learning without mixing: Towards a sharp analysis of linear system identification." In: *Proc. 31st Annual Conf. on Learning Theory.* 2018, pp. 439–473.

[92] T. R. V. Steentjes, M. Lazar, and P. M. V. den Hof. "Guaranteed $H_\infty$ performance analysis and controller synthesis for interconnected linear systems from noisy input-state data." In: *arXiv preprint arXiv:2103.14399* (2021).

[93] R. Strässer, J. Berberich, and F. Allgöwer. "Data-driven control of nonlinear systems: Beyond polynomial dynamics." In: *arXiv preprint arXiv:2011.11355* (2021).

[94] M. Tanemura and S.-i. Azuma. "Closed-loop data-driven estimation on passivity property." In: *Proc. IEEE Conf. Control Technology and Applications.* 2019, pp. 630–634.

[95] M. Tanemura and S.-i. Azuma. "Efficient data-driven estimation of passivity properties." In: *IEEE Control Systems Letters* 3.2 (2019), pp. 398–403.

[96] H.-D. Tran, L. V. Nguyen, W. Xiang, and T. T. Johnson. "Order-reduction abstractions for safety verification of high-dimensional linear systems." In: *Discrete Event Dyn. Syst.* 27 (2017), pp. 443–461.

[97] A. Tsiamis and G. J. Pappas. "Linear systems can be hard to learn." In: *arXiv preprint arXiv:2104.01120* (2021).

[98] S. Tu, R. Boczar, and B. Recht. "On the approximation of Toeplitz operators for nonparametric $\mathcal{H}_\infty$-norm estimation." In: *Proc. American Control Conf.* 2018, pp. 1867–1872.

[99] A. Van der Schaft. *$L_2$-gain and passivity techniques in nonlinear control.* Springer, 2000.

[100] J. Veenman, C. W. Scherer, and H. Köroğlu. "Robust stability and performance analysis based on integral quadratic constraints." In: *Eur. J. Control* 31 (2016), pp. 1–32.

[101] J. Veenman and C. W. Scherer. "IQC-synthesis with general dynamic multipliers." In: *Int. J. Robust Nonlinear Control* 24.17 (2013), pp. 3027–3056.

[102] H. J. van Waarde, M. K. Camlibel, and M. Mesbahi. "From noisy data to feedback controllers: Non-conservative design via a matrix S-lemma." In: *IEEE Trans. Automat. Control* 67.1 (2022), pp. 162–175.

[103] H. J. van Waarde, J. Eising, H. L. Trentelman, and M. K. Camlibel. "Data informativity: A new perspective on data-driven analysis and control." In: *IEEE Trans. Automat. Control* 65.11 (2020), pp. 4753–4768.

[104] H. J. van Waarde, C. de Persis, M. K. Camlibel, and P. Tesi. "Willems' fundamental lemma for state-space systems and its extension to multiple datasets." In: *IEEE Control Systems Letters* 4.3 (2020), pp. 602–607.

[105] B. Wahlberg, H. Hjalmarsson, and P. Stoica. "On estimation of the gain of a dynamical system." In: *Proc. 2011 Digital Signal Processing and Signal Processing Education Meeting*. 2011, pp. 364–369.

[106] B. Wahlberg, M. B. Syberg, and H. Hjalmarsson. "Non-parametric methods for $\mathcal{L}_2$-gain estimation using iterative experiments." In: *Automatica* 46.8 (2010), pp. 1376–1381.

[107] J. C. Willems. "Dissipative dynamical systems part I: General theory." In: *Arch. Ration. Mech. Anal.* 45 (1972), pp. 321–351.

[108] J. C. Willems, P. Rapisarda, I. Markovsky, and B. L. M. D. Moor. "A note on persistency of excitation." In: *Systems & Control Letters* 54.4 (2005), pp. 325–329.

[109] A. Xue and N. Matni. "Data-driven system level synthesis." In: *Proc. 3rd Conf. Learning for Dynamics and Control*. 2021, pp. 189–200.

[110] Y. Yan, J. Bao, and B. Huang. "Dissipativity analysis for linear systems in the behavioural framework." In: *Proc. Australian & New Zealand Control Conf.* 2019, pp. 152–156.

[111] H. Yang and S. Li. "A data-driven predictive controller design based on reduced Hankel matrix." In: *Proc. 10th Asian Control Conf.* 2015, pp. 1–7.

[112] H. Zakeri and P. J. Antsaklis. "A data-driven adaptive controller reconfiguration for fault mitigation: A passivity approach." In: *Proc. 27th Med. Conf. Control and Automation.* 2019, pp. 25–30.

[113] G. Zames. "On the input-output stability of time-varying nonlinear feedback systems part I: Conditions derived using concepts of loop gain, conicity, and positivity." In: *IEEE Trans. Automat. Control* 11.2 (1966), pp. 228–238.

[114] K. Zhou, J. C. Doyle, and K. Glover. *Robust and optimal control.* Prentice Hall, 1996.

## Publications of the Author

[AK1] A. Alanwar, A. Koch, F. Allgöwer, and K. H. Johansson. "Data-driven reachability analysis from noisy data." In: *arXiv preprint arXiv:2105.07229* (2021).

[AK2] A. Alanwar, A. Koch, F. Allgöwer, and K. H. Johansson. "Data-driven reachability analysis using matrix zonotopes." In: *Proc. 3rd Conf. Learning for Dynamics and Control*. 2021, pp. 163–175.

[AK3] J. Berberich, A. Koch, C. W. Scherer, and F. Allgöwer. "Robust data-driven state-feedback design." In: *Proc. American Control Conf.* 2020, pp. 1532–1538.

[AK4] A. Koch, J. Berberich, J. Köhler, and F. Allgöwer. "Determining optimal input-output properties: A data-driven approach." In: *Automatica* 134 (2021), p. 109906.

[AK5] A. Koch, J. Berberich, and F. Allgöwer. "Provably robust verification of dissipativity properties from data." In: *IEEE Trans. Automat. Control* (2021). early access.

[AK6] A. Koch, J. Berberich, and F. Allgöwer. "Verifying dissipativity properties from noise-corrupted input-state data." In: *Proc. 59th IEEE Conf. Decision and Control*. 2020, pp. 616–621.

[AK7] A. Koch, M. Lorenzen, P. Pauli, and F. Allgöwer. "Facilitating learning progress in a first control course via Matlab apps." In: *Proc. 21st IFAC World Congress*. 2020, pp. 17356–17361.

[AK8] A. Koch, J. M. Montenbruck, and F. Allgöwer. "Sampling strategies for data-driven inference of input-output system properties." In: *IEEE Trans. Automat. Control* 66.3 (2021), pp. 1144–1159.

[AK9] J. Köhler, L. Schwenkel, A. Koch, J. Berberich, P. Pauli, and F. Allgöwer. "Robust and optimal predictive control of the COVID-19 outbreak." In: *Annual Reviews in Control* 51 (2021), pp. 525–539.

[AK10] T. Martin, A. Koch, and F. Allgöwer. "Data-driven surrogate models for LTI systems via saddle-point dynamics." In: *Proc. 21st IFAC World Congress*. 2020, pp. 953–958.

[AK11] M. I. Müller, A. Koch, F. Allgöwer, and C. R. Rojas. "Data-driven input-passivity estimation using power iterations." In: *Proc. 19th IFAC Symp. System Identification*. 2021, pp. 619–624.

[AK12] P. Pauli, A. Koch, and F. Allgöwer. "Smartphone apps for learning progress and course revision." In: *Proc. 21st IFAC World Congress*. 2020, pp. 17368–17373.

[AK13] P. Pauli, A. Koch, J. Berberich, and F. Allgöwer. "Training robust neural networks using Lipschitz bounds." In: *IEEE Control Systems Letters* 6 (2022), pp. 121–126.

[AK14] P. Pauli, J. Köhler, J. Berberich, A. Koch, and F. Allgöwer. "Offset-free set-point tracking using neural network controllers." In: *Proc. 3rd Conf. Learning for Dynamics and Control*. 2020, pp. 992–1003.

[AK15] D. Persson, A. Koch, and F. Allgöwer. "Probabilistic H2-norm estimation via Gaussian process system identification." In: *Proc. 21st IFAC World Congress*. 2020, pp. 431–436.

[AK16] A. Romer, J. Berberich, J. Köhler, and F. Allgöwer. "One-shot verification of dissipativity properties from input-output data." In: *IEEE Control Systems Letters* 3.3 (2019), pp. 709–714.

[AK17] A. Romer, J.-Y. Kim, and L. J. Jacobs. "An analytical and numerical study of the nonlinear reflection at a stress-free surface." In: *AIP Conf. Proc.* 2015, pp. 1616–1623.

[AK18] A. Romer, J.-Y. Kim, J. Qu, and L. J. Jacobs. "The second harmonic generation in reflection mode: An analytical, numerical and experimental study." In: *J. Nondestructive Evaluation* 35.6 (2016).

[AK19] A. Romer, J. M. Montenbruck, and F. Allgöwer. "Data-driven inference of conic relations via saddle-point dynamics." In: *Proc. 9th IFAC Symp. Robust Control Design*. 2018, pp. 586–591.

[AK20] A. Romer, J. M. Montenbruck, and F. Allgöwer. "Determining dissipation inequalities from input-output samples." In: *Proc. 20th IFAC World Congress*. 2017, pp. 7789–7794.

[AK21] A. Romer, J. M. Montenbruck, and F. Allgöwer. "Sampling strategies for data-driven inference of passivity properties." In: *Proc. 56th IEEE Conf. Decision and Control*. 2017, pp. 6389–6394.

[AK22] A. Romer, J. M. Montenbruck, and F. Allgöwer. "Some ideas on sampling strategies for data-driven inference of passivity properties for MIMO systems." In: *Proc. American Control Conf.* 2018, pp. 6094–6100.

[AK23] A. Romer, S. Trimpe, and F. Allgöwer. "Data-driven inference of passivity properties via Gaussian process optimization." In: *Proc. European Control Conf.* 2019, pp. 29–35.

[AK24] M. Sharf, A. Koch, D. Zelazo, and F. Allgöwer. "Model-free practical cooperative control for diffusively coupled systems." In: *IEEE Trans. Automat. Control* (2021). early access.

[AK25] C. Verhoek, R. Tóth, S. Haesaert, and A. Koch. "Fundamental lemma for data-driven analysis of linear parameter-varying systems." In: *arXiv preprint arXiv:2103.16171* (2021).

[AK26] J. Vinogradska, B. Bischoff, D. Nguyen-Tuong, H. Schmidt, A. Romer, and J. Peters. "Stability of controllers for Gaussian process forward models." In: *Proc. 33rd Int. Conf. Machine Learning*. 2016, pp. 545–554.

[AK27] N. Wieler, J. Berberich, A. Koch, and F. Allgöwer. "Data-driven controller design via finite-horizon dissipativity." In: *Proc. 3rd Conf. Learning for Dynamics and Control*. 2021, pp. 287–298.